IMAGES
of America

MARSHALL SPACE FLIGHT CENTER

By the mid-1960s, this sign, standing at an unknown location, said it all: Huntsville was not only the home of the US Army Redstone Arsenal, but also the National Aeronautics and Space Administration's (NASA) George C. Marshall Space Flight Center. The back of the postcard this sign was featured on proclaims, "Welcome to the 'Outer Space Capital, USA'—a progressive and friendly metropolis." (Author's collection.)

ON THE COVER: Technicians at Marshall Space Flight Center hoist a dynamic test version of the Saturn IB's second stage (S-IVB) into Building 4557, Dynamic Test Stand (Saturn I/IB), in the East Test Area. The S-IVB, with a single J-2 engine, was built by the Douglas Aircraft Company in St. Louis, Missouri. A fully assembled Saturn IB was tested for structural soundness on January 18, 1965. (Courtesy of NASA.)

IMAGES
of America

MARSHALL SPACE FLIGHT CENTER

Cindy Donze Manto
Foreword by Wanda A. Sigur,
Lockheed Martin Civil Space
Vice President (Ret.)

ARCADIA
PUBLISHING

Published by Arcadia Publishing
Charleston, South Carolina

Printed in the United States of America

Library of Congress Control Number: 2019949617

For all general information, please contact Arcadia Publishing:
Telephone 843-853-2070
Fax 843-853-0044
E-mail sales@arcadiapublishing.com
For customer service and orders:
Toll-Free 1-888-313-2665

Visit us on the Internet at www.arcadiapublishing.com

To my husband, Fulvio Manto, an aerospace engineer who immigrated from Italy and began his career at Michoud Assembly Facility and worked at Marshall Space Flight Center.

CONTENTS

FOREWORD

Rarely is "doing the impossible" on the top of a to-do list. Whether transitioning ballistic missiles to man-rated launch vehicles, building rockets to send men to the Moon or Mars, or providing the propulsion system for the space shuttle, the challenges faced by Marshall Space Flight Center (MSFC) have been earth-shattering—or at least earth-departing.

From nascent advanced composites engineer to the space shuttle external tank (ET) program manager, my decades of engagements with MSFC have populated so many stories about the people of Marshall and their dedication to these challenges and to their teams.

Like many engineers, my early career involved testing—lots of it. In the middle of that effort, I was given the gift of context. NASA's materials & processes veteran, Dr. Jim Stuckey, provided that little catalyst to a new engineer by explaining the importance of those oxygen compatibility tests through the context of the tragedy of the decision to use pressurized pure oxygen in the crew cabin on Apollo 1. With context came the perspective that triggered a sense of purpose that forever tied me to a bigger mission.

The Marshall legacy is a collection of stories of leadership. I recall the return-to-flight campaign of the space shuttle external tank following the *Columbia* accident, when suddenly, everything became more complicated. While developing the solution to foam loss, knee-deep in the complex design space of configuration, process, cryo-pumping, and ascent environments, Hurricane Katrina hit the city of New Orleans. With 96 percent of the workforce affected, the return to production of the Michoud Assembly Facility was halted. With tight recovery timelines, John "Chap" Chapman, NASA ET project manager, chose to lead his team in gutting the flooded homes of a devastated workforce. Today, leadership examples extend to MSFC working with nascent NewSpace companies and our next generation of space pioneers.

Whether your interests are rocket boosters, engines, payloads, testing, or just big space missions, this book will prove exciting reading. I had the great fortune to meet Cindy Manto years ago through her husband, Fulvio, then a stress analyst on ET. Having had a front row seat to the action, Cindy tells the stories of MSFC with the insight and background of an insider. She has captured the stories of the many campaigns of Marshall and those who have made our nation's space legacy a reality by accepting the challenge to do the impossible.

—Wanda Sigur
Lockheed Martin Civil Space Vice President (Ret.)

ACKNOWLEDGMENTS

Special thanks to Wanda A. Sigur; Shalis Worthy, archivist, Huntsville–Madison County Public Library, Huntsville, Alabama; Meredith McDonough, digital assets coordinator, Alabama Department of Archives and History, Montgomery, Alabama; Lance N. George, Keller Archives, Huntsville; Rhett Breerwood, historian, the Jackson Barracks Military Museum, New Orleans, Louisiana; Richard Moran, curator, Louisiana Maneuvers and Military Museum at Camp Beauregard, Pineville, Louisiana; Matthew C. Hansen, archivist, Franklin D. Roosevelt Presidential Library and Museum, Hyde Park, New York; Janice Davis, archives technician, Harry S. Truman Presidential Library and Museum, Independence, Missouri; Mary Burtzloff, audiovisual archivist, Eisenhower Presidential Library and Museum, Abilene, Kansas; Maryrose Grossman, audio-visual reference, John F. Kennedy Presidential Library and Museum, Boston, Massachusetts; Heather Moore, photograph historian, US Senate History Office, Washington, DC; and Paul Barron, director of libraries and archives, the George C. Marshall Foundation, Lexington, Virginia.

Many thanks to NASA; the Library of Congress; University of North Texas Libraries; State Archives of North Carolina; Kelly Kazek, regional reporter/columnist for Alabama Media Group, AL.com; Fulvio Manto, director of mechanical and propulsion design (retired), Lockheed Martin Space Systems; Kenneth J. Donze, for his expertise in automobiles; and Mike Jetzer of heroicrelics.org.

Please note that this is not an official publication of NASA and that any opinions expressed or errors in content are the responsibility of the author.

INTRODUCTION

Marshall Space Flight Center is NASA's largest field center devoted to propulsion analysis and development. Its origins lie in the Army Ballistic Missile Agency (ABMA) headquartered at its Redstone Arsenal in Huntsville, Alabama. The ABMA was established at the height of the Cold War missile race of the late 1940s through the 1950s and was staffed primarily with German engineers and scientists brought to the United States at the end of World War II via the US Army's Operation Paperclip.

During the final months of World War II, Army officials began advocating for the exploitation of German scientific knowledge, particularly of the V-1 and V-2 weapons, before the possible Soviet seizure of the former Nazi fabrication plants and potential hostage taking of scientists, engineers, and their families. Maj. Gen. Holger N. Toftoy asserted that "there is no quicker way to stimulate use of a new weapon than to discover it in use by the enemy." In 1945, he requested the immediate transport to the United States of 300 scientists and engineers from Germany. Receiving no answer, he flew to Washington himself to expedite action, earning the nickname "Mr. Missile," and to initiate Operation Paperclip.

Originally code-named Operation Overcast, the application process for transfer to the United States, via the State Department, had become compromised when the families began calling their US military housing "Camp Overcast." Army intelligence officers who were reviewing security reports of certain scientists began to discreetly attach a paperclip to the files of the more dubious cases, creating a new code name.

Initially, Dr. Wernher von Braun and seven V-2 specialists arrived at Fort Strong, Massachusetts, on September 20, 1945, to process their own documents, 18 days after the end of World War II on September 2. They were then transferred to Fort Bliss, Texas, and White Sands Proving Ground (WSPG), New Mexico, on December 10, 1945, joined by 55 more German émigré specialists working closely with Army Ordnance and General Electric personnel. By May 18, 1948, Operation Paperclip had brought a total of 1,136 German émigrés to the United States along with 100 nearly complete V-2s and 300 railroad boxcars of V-2 components, plans, manuals, and other documents.

Fort Bliss Research and Development and WSPG were the US centers for rocket development until 1950. In 1949, the Army Ordnance Department launched America's first sounding rocket, the WAC Corporal, which set an altitude record of 50 miles (space starts at 60 miles), which stood for the next eight years. A classified program, its name is widely believed to stand for "Without Attitude Control," because its fins were its only means of control.

In November 1950, a total of 500 military personnel, 130 German scientists and engineers, 180 General Electric contractor personnel, and 120 civil service employees moved to Redstone Arsenal. Between 1952 and 1953, the workhorse of the arsenal was developed, the Redstone ballistic missile. At Redstone, the feasibility of applying rocket propulsion to spaceflight was proposed by Dr. von Braun, called Project Orbiter, a concept for launching an Earth satellite. It was rejected, but its designs and hardware were utilized in the missile program for testing reentry nose cones.

Since ballistic missiles were now deemed to be a viable component of national defense, the ABMA was established on February 1, 1956. Research and development continued unabated up to 1957, when the Soviet Union successfully launched Sputnik, Earth's first artificial satellite. In turn, the Army's "space team" successfully launched America's first satellite, Explorer 1, with scientific data-gathering instruments aboard, in January 1958, three months after authorization. The launch was achieved with a modified Jupiter-C (Juno) launch vehicle (a four-stage system named Juno I), a derivative of the Redstone ballistic missile, itself a direct descendant of the V-2 rocket.

The following year, Pres. Dwight D. Eisenhower created the National Aeronautics and Space Administration on July 20, 1958, which would support a vigorous civilian space program, including existing laboratories and installations.

The Juno program spurred development of a large booster program for advanced space missions. In 1959, the Advanced Research Projects Agency (ARPA) authorized ABMA to begin research and development for a vehicle with 1,500,000 pounds of thrust by clustering eight rocket engines into one stage. The Juno program was renamed Saturn and received the highest national priority rating.

On July 1, 1960, the George C. Marshall Space Flight Center came into being with the transfer of 4,670 civil service employees, 1,840 acres of arsenal property, and facilities worth $100 million. On September 8, 1960, the center was officially dedicated to Gen. George C. Marshall in the heart of Redstone Arsenal.

President Eisenhower named the new field center, a NASA center for excellence in propulsion design, after George C. Marshall, a noted general of the Army, statesman, and Nobel Peace Prize recipient as the architect of the Marshall Plan immediately following World War II. In his speech, the president cited how Marshall's wartime experience enabled him to appreciate the problems of peace stemming from the impact of scientific and technological developments.

As the United States entered the Cold War space race, MSFC's primary responsibility was the development of the Saturn family of launch vehicles used in the Apollo unmanned and manned lunar-landing program. Under the direction of von Braun, a series of unique facilities began to populate the landscape at MSFC including the Dynamic Test Stand, Propulsion and Structural Test Facility, and the Neutral Buoyancy Simulator (NBS).

MSFC developed and tested the F-1 engine, the largest, most powerful liquid rocket engine ever developed in the free world. It was responsible for the Skylab experimental space program, the Apollo-Soyuz Test Project, and the Apollo Telescope Mount. It also included the three major components of the space shuttle: the solid rocket booster (SRB), the external tank (ET), and the orbiter's space shuttle main engines (SSMEs).

Its achievements are numerous, but its unqualified success was landing the first human on the Moon, eight years and two months after announcing the intention to do so. Its other unqualified success was building and continuing to maintain the International Space Station, an orbiting laboratory for sustained science on an international basis.

Today, MSFC has a substantial role in the development of project Artemis as the United States prepares to return to the Moon and beyond. In addition to Artemis, the center continues to focus on the development of transportation and propulsion systems, space infrastructure, applied materials and manufacturing processes, and scientific research and instruments.

In 1970, the state of Alabama established the US Space and Rocket Center, which serves as an official NASA visitor center and a museum of manned and unmanned US spaceflight hardware. During the 1980s, five MSFC sites were designated national historic landmarks: the Redstone Test Stand, Propulsion and Structural Test Facility, Dynamic Test Stand (Saturn I/IB), Neutral Buoyancy Simulator, and the Saturn V display. The designation automatically lists these historic relics in the National Register of Historic Places, extending safeguards and benefits provided by federal law.

Meanwhile, the city of Huntsville, whose economy was originally based primarily on the cultivation of cotton before transitioning to textile mills and attempting to diversify its agriculture with watercress prior to World War II, has been propelled onto the national and world stage as a center of rocketry and propulsion.

One

KING COTTON AND MADISON COUNTY

1808–1899

In 1798, most of present-day Alabama became part of the Mississippi Territory, which stretched from the state of Georgia to the Mississippi River. In 1805, John Hunt, a veteran of the American Revolutionary War with Great Britain, arrived at Big Spring and established the town of Hunt's Spring. Native Americans, mainly the Chickasaw and Cherokee tribes, who had occupied the surrounding fertile ground bordered by the Tennessee River to its south for several centuries, saw their land sold from under them.

Mississippi Territory governor Robert Williams created Madison County, named for Pres. Thomas Jefferson's secretary of state, James Madison, on December 13, 1808. In 1809, LeRoy Pope, a planter from Georgia, bought 60 acres of land around Big Spring at a public auction, naming it Twickenham after his British ancestral town. By 1809, it was renamed Huntsville, in response to the prevalent anti-British sentiment as that country struggled to maintain a presence in North America. That struggle ended with the nearby Battle of New Orleans in 1815.

Rapid growth by 1816 brought a two-story brick courthouse surrounded by the 60 acres of land, divided into two-acre blocks, each subdivided into four lots. Cotton growing, ginning, and milling, mainly by use of slave labor, brought great wealth that supported a broadening commercial economy. On December 14, 1819, Alabama attained statehood, and Madison County became the cotton trading center of the Tennessee Valley through the 1850s. By 1855, the Madison & Charleston Railroad established the first link between the Atlantic coast and the lower Mississippi River.

William Frye, a German immigrant from Bavaria, painted this bucolic scene of Huntsville with its centrally located Big Spring in the 1850s. The carriage rider at right would be traveling north on present-day Church Street. Named after its builder in 1825, the Fearn Canal at center carried cotton from Big Spring to the Tennessee River at the port of Triana. (Courtesy of the Huntsville–Madison County Public Library.)

This 1859 view down Jefferson Street, facing south, shows the Huntsville Hotel (left), the city hall (center), and a two-story Italianate commercial structure that was replaced in 1877 by the block-long McGee Hotel. The hotel sought to imitate the Greek Revival American Townhouse style prevalent in New Orleans's French Quarter. (Courtesy of the Huntsville–Madison County Public Library.)

In 1864, Union troops occupied the most visible and central location in Huntsville, Courthouse Square. Reportedly, the soldiers tore down North Alabama College, using its bricks to construct their winter quarters. The Greek Revival–style Northern Bank of Alabama is visible at far left. Today, the building stands and operates as a Regions Bank. (Courtesy of the Huntsville–Madison County Public Library.)

Looking east from Courthouse Square is a view down Eustis Avenue in 1865. The Church of the Nativity (center), an Episcopal congregation, was established in 1843 and continues to thrive in that location. The present church, one of the finest examples of Gothic architecture in the United States, was built in 1859. The original church was razed in 1878. (Courtesy of Huntsville–Madison County Public Library.)

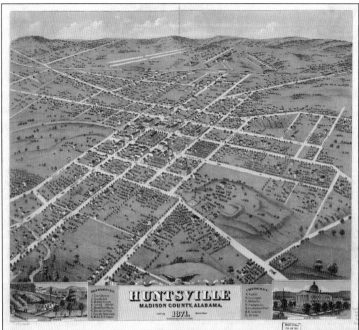

Following the Civil War, this Reconstruction-era map from 1871 shows the city of Huntsville clearly intact. The Madison County Courthouse (left center), facing Eustis Street, is the center of town with an array of churches within one to three blocks; Big Spring Waterworks is one block away to the left, and the Memphis & Charleston Railroad skirts the edge of town. (Courtesy of the Library of Congress.)

The Bell Mill, built in 1820 and rebuilt in 1841 after a fire, was Alabama's first spinning and weaving cotton textile plant, with 3,000 spindles and 100 looms. Power was generated by a waterwheel on the dammed Flint River and then by steam plant by 1868. The pre–Civil War slave cabins gave way to a small village until 1885, when the mill closed. (Courtesy of the Huntsville–Madison County Public Library.)

Fieldworkers are picking cotton in Madison County. Before the invention of the cotton gin (or engine), it took over 600 hours of labor to produce one bale of cotton, making large-scale production uneconomical even with unpaid slave labor. Eli Whitney's cotton gin reduced that to approximately 12 hours per bale. American cotton exports declined sharply during the Civil War but easily rebounded by 1865. (Courtesy of the Huntsville–Madison County Public Library.)

Wagonloads of cotton continued to be gathered in downtown Huntsville throughout the Civil War, including during two separate occupations by Union troops: for a few months in 1862, and from late 1863 to the end of 1864. Cotton continued to serve as the economic engine for Huntsville during Reconstruction, and helped start the local textile industry. The commercial building at right housed a "cash store," specializing in specie and banknotes, and a dry goods store. (Courtesy of Alabama and Department of Archives and History.)

Postwar prosperity brought county fairs and entertainment to the towns and surrounding counties. A county fair band poses for a photograph alongside the county courthouse in 1887. From left to right are C.F. Bost, tuba; Tom Weaver, baritone; Charles Varin, bass drum; H.G. Poor, tenor trombone; Louis Dinnigan, snare drum; Charles Conkel, alto; "Bud" Erwin, alto; John L. Hay, cornet; and D.C. Monroe, cornet. (Courtesy of the Huntsville–Madison County Public Library.)

Mule-drawn wagons of cotton vie for the best sales location along the north side of Courthouse Square in downtown Huntsville in 1899. The seat of Madison County, Huntsville was also the center of cotton production and sales. On the left, two trolley cars travel the east-west Randolph Street route; on the right stands Madison County's second courthouse on the site, designed by architect George Steele in 1838. (Courtesy of Huntsville–Madison County Public Library.)

In 1898, buffalo soldiers of the 10th US Cavalry assemble at Cavalry Hills in northwest Huntsville, near University Drive and Pulaski Pike, after the Spanish-American War. They fought in every major conflict from the Spanish-American War until 1952, when Pres. Harry S. Truman integrated the unit with the regular Army. (Courtesy of the Huntsville–Madison County Public Library.)

On horseback, Gen. A.K. Arnold (left) and Gen. Joseph Wheeler review the 4th Army Corps (Spanish-American War) on parade on the north side of Courthouse Square in downtown Huntsville in 1898. General Wheeler served as a cavalry general in the Confederate army and as general in the US Army. The 4th Army Corps was stationed at Huntsville and disbanded on January 16, 1899. (Courtesy of the Huntsville–Madison County Public Library.)

Veterans of the Spanish-American War gather outside the Huntsville Hotel in 1899 while recuperating from wounds and diseases contracted during their tour of duty. The war between Spain and the United States arose when an internal explosion aboard the USS *Maine* in Havana harbor was attributed to sabotage during the Cuban War of Independence. (Courtesy of the Huntsville–Madison County Public Library.)

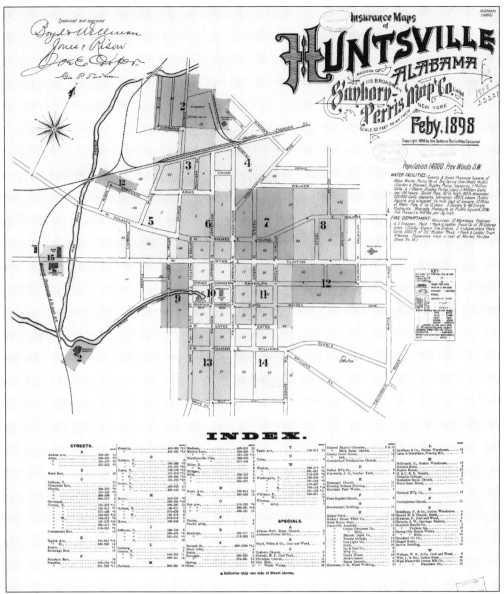

This 1898 Sanborn insurance map of Huntsville displays an orderly grid pattern with Courthouse Square in the center. On the right are inclusive descriptions of the area's waterworks and fire department, both urban amenities. Note that the expansive rail connections now included the Nashville, Chattanooga & St. Louis Railroad. (Courtesy of the Library of Congress.)

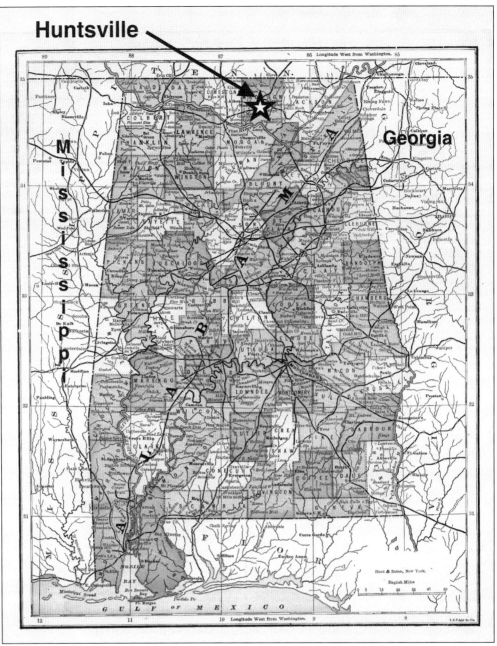

This 1893 map of Alabama shows Huntsville's location relative to the rest of the state and the Gulf of Mexico, 450 miles south. With its location on the Tennessee River connecting it to the Mississippi River via the Ohio River, along with the advent of railroads for freight and passenger travel, Huntsville was poised for growth in the 20th century. (Courtesy of the Huntsville–Madison County Public Library.)

Two

INDUSTRIALIZATION AND HUNTSVILLE
1900–1939

Industrialization arrived with the 20th century in the form of immense textile mills. The mills mirrored the growth of Huntsville itself as it constructed infrastructure to support the mills, the workers, and all the necessary and incidental businesses a growing city required. Drawing on the county's prodigious cotton cultivation, accessible water and railroad transportation, relatively mild climate, and available land for development, industrialists and investors, mainly from the Northeast, descended on the region to build some of the largest mills in the United States.

The Dallas Mill, built in 1892 on 50 acres of land in northeast Huntsville, was quickly followed by the Lowe Mill on Seminole Drive and Ninth Avenue in 1900, the Merrimack Mill a half mile from Brahan Spring also in 1900, and the Lincoln Mill on Oakwood Avenue in 1918. At their peak in 1900, the mills together employed approximately 5,000 men, women, and children from a population of over 8,000.

Without safety regulations or child labor laws, and with meager education requirements, families lied about their children's ages: some as young as 8 to 10 years old would be passed off as 12. State law required eight weeks of school: six had to be consecutive weeks, for six hours per day, starting at six in the morning, and then work in the mill at noon. This schedule was reversed to accommodate children on the morning shift. Federal law finally set minimum work standards for children in 1938.

The mills endured the Great Depression by lowering pay and adding hours, leading to the multistate textile workers' strike of 1934, from which they never fully recovered.

Developing concurrently with the mills were a series of wholesale plant nurseries, including Chase, Naugher, Byers, E.F. Dubose, and Webb. For a brief time, Huntsville became the "watercress capital of the world" during the 1930s, thanks to the Dennis Watercress Company, which specialized in the mass cultivation of the native vegetable.

On the eve of World War II, a representative of the fifth congressional district and two local businessmen thought Huntsville's economy should be further diversified.

By 1899, Huntsvillians were celebrating the Fourth of July with parades led by highly decorated carriages. From left to right are Robert Weedon, Sarah Dement, Mamie Fletcher, and Hector Lane. (Courtesy of Alabama Department of Archives and History.)

Founded in 1887 by Michael Cudahy, with backing from Philip D. Armour in Omaha, Nebraska, the Cudahy Packing Co. was one of the largest packing houses in the United States by 1922. Because of its location on the east-west railroad lines between the Atlantic coast and the Mississippi River, Huntsville was one of 97 cities around the country with distribution operations for the company. (Courtesy of Huntsville–Madison County Public Library.)

The Dallas Mill, built by Trevanion B. Dallas of Nashville, Tennessee, began operations in 1892 as Alabama's largest mill producing cotton sheeting and woolen goods. Its employee village, extending from Oakwood Avenue to Dallas Street, included homes, a medical clinic, churches, a library, YMCA, and schools. After closing in 1949, the city incorporated the village in 1955. The unoccupied building burned down in 1991. (Courtesy of the Huntsville–Madison County Public Library.)

A traction engine arrives in downtown Huntsville on Randolph Street, along the north side of Courthouse Square, sometime between 1900 and 1909. Used for rugged outdoor projects, traction engines burned coal, wood, or straw and used about 300 gallons of water to generate steam that allowed the engine to plow, haul, and pull as much as 55,000 pounds. (Courtesy of Alabama Department of Archives and History.)

Two employees are working the roving machines at the Merrimack Mill in the late 1930s. "Roving" is a smaller strand of cotton that has been further reduced by rollers and bobbin speed. To resist breaking, the roving is given many twists as it goes on the bobbin. Opened on July 9, 1900, the mill became the Huntsville Manufacturing Company on January 14, 1946. (Courtesy of the Huntsville–Madison County Public Library.)

The equipment in the weaving room at the Merrimack Mill receives routine maintenance during a shift. Weaving transforms yarn into cloth by interlacing two sets of threads in the machine or loom. The set of yarn that runs the length of the goods is called the "warp," and the yarn that runs at right angles to the warp is the "filling." (Courtesy of the Huntsville–Madison County Public Library.)

The Merrimack Manufacturing Company, based in Lowell, Massachusetts, constructed a mill in Huntsville in 1900, followed by a second mill in 1904. The mammoth complex consisted of 90,000 spindles and 2,900 looms. At its peak in 1955, it had 1,600 employees, 145,896 spindles, and 3,437 looms. Huntsville's last mill closed in 1988 and was demolished in 1992. It is now Huntsville Park, near Triana Boulevard SW. (Courtesy of the Huntsville–Madison County Public Library.)

Employees of the Merrimack Mill sit for a photograph in the early 1900s. Children as young as eight years old worked as sweepers and doffers. A doffer removes, or doffs, bobbins or spindles holding spun fiber from a spinning frame and replaces them with empty bobbins and spindles. (Courtesy of the Huntsville–Madison County Public Library.)

Twelve-year-old Eliza, eight-year-old Pinkie Durham (his mother insisted that he was 12), and an unidentified sibling behind them are pictured at home in 1913. In the absence of child labor laws, both worked for the Merrimack Manufacturing Company. Pinkie worked as a sweeper, while Eliza was recovering from a broken leg when a boy ran a doffing box into her at the mill. (Courtesy of the Library of Congress.)

Every mode of transportation—horses and buggies, bicycles, automobiles, and elephants—is on display as the circus winds its way around the west side of Courthouse Square in 1913. The automobile at lower right has its steering wheel on the right. After 1908, the Ford Motor Company's Model T changed the wheel to the left, making it easier for passengers to enter the car while avoiding traffic. (Courtesy of Huntsville–Madison County Public Library.)

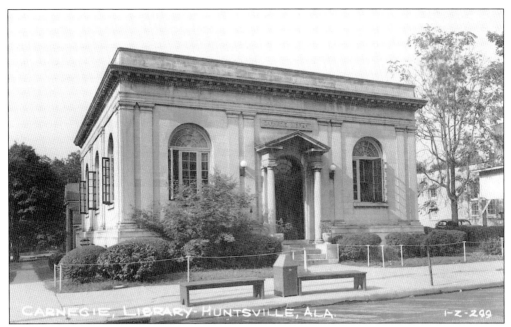

After receiving a grant from Andrew Carnegie, Huntsville's Carnegie Library, located at the corner of Madison Street and Gates Avenue, was built in 1915 and designed in the "Carnegie Classical" style. These libraries were commonly seen as an award of cultural merit to a deserving community. It was torn down in 1966 and replaced with a parking lot, the victim of urban renewal. (Courtesy of the Huntsville–Madison County Public Library.)

Pictured in downtown Huntsville are three types of transportation: a Perley Thomas streetcar, a horse and buggy, and around the corner, an early Model T Ford. The streetcar sports a safety net in front to prevent people and animals from falling beneath its wheels. Until mass production reduced the cost of automobiles, buggies continued to be the common means of local transportation, costing between $25 and $50. (Courtesy of Huntsville–Madison County Public Library.)

Twelve Confederate veterans of the Civil War gathered at the corner of Franklin and Gates Streets for a reunion in 1938. From left to right are (first row) J.K.P. Ketchum, G.W. Chumly, Gen. Paul Sanguinetti, Dr. C.C. Jones, J.S. Thompson, J.W. Comb, Col. H.M. Bell, Gen. J.R.K. Snody, Gen. R.T. Boatright, and J.W. Dixon; (second row) Rev. R.A. Gwen and Col. Simon Phillips. (Courtesy of the Huntsville–Madison County Public Library.)

In an effort to diversify the local economy during the Great Depression of the 1930s, some entrepreneurs invested in watercress as a hedge against declining cotton prices. Watercress already had an affinity for the soil, climate, and slow moving bodies of water in the area when Frank Dennis of New Jersey discovered that northern Alabama was ideal for year-round cultivation in 1874. (Courtesy of the Huntsville–Madison County Public Library.)

At the Huntsville Depot, workers wait with barrels of locally grown watercress to be loaded onto a freight train destined for dining tables in far-flung cities such as New York, New Orleans, and Chicago during the 1930s. Dennis Watercress Company of Huntsville was the largest producer, growing and harvesting an estimated two million bunches in the area from the early 20th century through the 1960s. (Courtesy of the Huntsville–Madison County Public Library.)

Three

WORLD WAR II AND REDSTONE ARSENAL
1940–1946

On the eve of America's entry into World War II, the Chemical Warfare Service was searching for another site to supplement its main location at Edgewood Arsenal, Maryland. Although the United States was a proponent of the Geneva Protocol of 1925 that sought to prohibit the export of gases for use in war, it felt that this prohibition would cease to apply to them if the prohibited weapons were used against them. The US Senate was heavily lobbied by the US military and the American Chemical Society against ratification. Before World War II, many of the great powers ratified the protocol except the United States and Japan. The Senate finally ratified it in 1975.

Two prominent Huntsville businessmen, Lawrence Goldsmith and George Mahoney, and US representative John J. Sparkman, convinced the Army that Huntsville offered the right combination of weather, water, electrical power, transportation in the form of railroads, and an inland location for greater security. Over 550 families, mostly tenants and sharecroppers, were displaced from about 30,000 acres. A land use agreement between the Army and the Tennessee Valley Authority (newly established in 1933 and supplying electrical power to the region) added another 1,250 acres along the Tennessee River.

The new military installation was comprised of the Huntsville Arsenal and Huntsville Depot, both operated by the Chemical Warfare Service, and the Redstone Ordnance Plant (renamed Redstone Arsenal in 1943), operated by the Army Ordnance Department. Both won numerous Army-Navy "E" Awards for excellence in production during wartime operations.

Huntsville Arsenal produced and stockpiled chemical weapons such as phosgene gas, Lewisite, mustard gas, tear gas, and carbonyl iron powders for radio and radar tuning. The Army Ordnance Department produced smoke and incendiary devices and pyrotechnic devices including small solid-fuel rockets.

However, a mere three days after Japan's surrender on September 2, 1945, production ceased. The average workforce of 4,400 dropped to 200-250 civilian employees, the land reverted to agriculture, and new tenants were sought to occupy the buildings. By 1947, the installation was declared surplus and advertised for sale.

In 1940, US representative John J. Sparkman, along with Huntsville businessmen Lawrence Goldsmith and George Mahoney, persuaded the federal government to locate the $40-million arsenals to Huntsville as a way to help diversify the local economy of mainly cotton production. Later, as senator, Sparkman was influential in establishing missile and rocket research at the Redstone Arsenal. (Courtesy of the US Senate Historical Office.)

The Merrimack Manufacturing Company was built in 1900 and boasted its own village, hospital, school, and about 279 houses for employees, among other amenities. Although ravaged by the Great Depression and strikes during the 1930s, business rebounded greatly during World War II as the demand for cotton increased, along with demand for housing in Huntsville. It became Huntsville Manufacturing Company in 1947, and its village is now a historic district. (Author's collection.)

The gala Cotton Ball admonished everyone to "Buy More Cotton!" in the early 1940s. Dormant cotton prices throughout the Great Depression steadily increased as textile mills won contracts to produce uniforms, bedding, tents, and sandbags. Some farmers relied on prisoners of war to chop cotton, harvest crops, and fell timber. Nearly 4,000 prisoners helped to save Alabama's $38 million peanut crop in 1944. (Courtesy of the Huntsville–Madison County Public Library.)

Over the cotton fields surrounding Huntsville flies a North American B-25 Mitchell, a medium bomber that generally carried about two tons of bombs. The plane climbs as cluster bombs skip across the target mat, 500 feet in diameter, toward the 75-foot-wide-by-25-foot-high proving wall. Government censors allowed this photograph and information to be made public by October 1945. (Author's collection.)

The intelligence division of Huntsville Arsenal had "no objection to this photograph" in October 1945. What may appear to be cotton, ready to be picked, is in reality skip bomb testing in the open fields around the arsenal where practice runs to strategically drop M-50 incendiary bombs became routine. The Chemical Warfare Service was testing $21.5 million worth of munitions in a five-month period. (Author's collection.)

A B-25 Mitchell practices target bombing at Huntsville Arsenal on February 20, 1945, near the end of World War II. The B-25 ascends rapidly with bomb bays open after releasing an M-50 bomb from an altitude of 70 feet at 250 miles per hour. The 100-pound cluster breaks in midair into individual stick-like incendiaries, striking the reinforced concrete bombing mat and proving wall at an angle with tremendous force. (Author's collection.)

Mock residential housing was used to conduct the experimental firebombing of cities and towns in Nazi Germany and Japan. This structure at Redstone Arsenal appears to have used basic building techniques, representing a generic residential structure. Other "villages," built by the Army in different locations, used authentic construction techniques and materials in order to find a tactic to achieve a firestorm in city centers. (Author's collection.)

In June 1943, a soldier fills "a cylinder with freshly combined war gas" that was "made at an eastern arsenal." It was probably Edgewood Arsenal adjacent to Aberdeen Proving Ground in Harford County, Maryland. The press release assured the American public that "if, and when, the Axis may decide to resort to the use of poison gas against the Allies, the United States is well prepared to retaliate in kind." (Author's collection.)

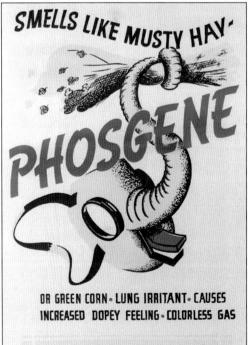

SMELLS LIKE MUSTY HAY

PHOSGENE

OR GREEN CORN • LUNG IRRITANT • CAUSES INCREASED DOPEY FEELING • COLORLESS GAS

The sign to the right reads, "Section 1," while a B-29 heavy bomber, which excelled at both high-altitude strategic bombing and low-altitude night incendiary bombing, goes through its practice runs at Dugway Proving Ground, about 80 miles outside of Salt Lake City, in 1943. Established in 1943 when the US Army Chemical Warfare Service required a more remote location to conduct tests, its activities remain a closely guarded secret. (Author's collection.)

Colorful posters with cartoon drawings and clever verse were used to strongly warn the troops of the dangers of the various types of gases that could be used in the event of chemical warfare. This poster described the deceptive aroma of one of the most common, phosgene, which is a chlorine gas. Huntsville Arsenal produced 20,000 tons of phosgene, along with lewisite and Levinstein mustard. (Courtesy of the University of North Texas Libraries.)

Early in World War II, Pres. Franklin D. Roosevelt established a no-first-use policy for chemical weapons. However, Building 481 was devoted to the manufacture of gas masks. Postwar, a disassembly line was created to recycle the parts into new lightweight masks, producing about 4,000 masks per eight-hour shift. In September 1946, the gas mask assembly plant ceased operations. (Courtesy of the University of North Texas Libraries.)

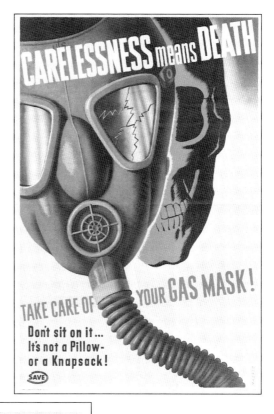

In 1943, the US Office of Civil Defense issued an extensive chemical warfare reference and training chart for use by the general population. Despite their prohibition after World War I, nations continued to stockpile them. Chemical weapons were used against civilians in concentration camps, but neither the Axis nor Allied powers deployed them in combat. (Courtesy of the State Archives of North Carolina.)

Two technicians use a press to form the bottom of a shell casing for a 105-millimeter artillery round at the Redstone Arsenal, which was operated by the US Army Ordnance Department in 1944. The hole or indentation in the center of the shell casing is where the ignitor cap would be placed. Redstone Arsenal also manufactured smoke and incendiary devices for the war effort. (Courtesy of the Huntsville–Madison County Public Library.)

On April 6, 1942, the city of Huntsville and Redstone Ordnance Plant participated for the first time in the national celebration of Army Day with a parade around Courthouse Square. Since April 6 is the anniversary of the United States' entry into World War I, the intent is to acquaint the public with the importance of national defense and Army activities, such as the Chemical Warfare Service. (Courtesy of the Huntsville–Madison County Public Library.)

The New 1948 KELLER Super Chief and Chief AERO ENGINEERED AUTO for Tomorrow

Following the end of World War II, George D. Keller, a former executive with Chrysler Corporation, Packard, and Studebaker, envisioned "a revolution" in public demand for basic automobiles that got 35 miles to the gallon at a sticker price of $750–$1,000. Since the established Detroit automakers remained uninterested in inexpensive, basic transportation, Keller competed with start-up automakers Davis, Tucker, and Playboy. (Courtesy of Lance George/Keller Archives.)

Days after the end of World War II on September 2, 1945, war production ceased. Huntsville and Redstone Arsenals, along with the Huntsville Depot, were declared surplus and advertised for sale. In 1948, the Keller Motor Corporation manufactured 18 automobiles before the sudden death of its founder, George Keller, resulted in its closure. Inside Building 481, the McVays, owners of a dealership in Kansas City, Missouri, pose with a Keller. (Courtesy of Lance George/Keller Archives.)

Four

THE COLD WAR
MISSILE RACE
1946–1960

As the outcome of World War II was becoming clear, US Army officials began advocating for the exploitation of German scientific knowledge, particularly of the V-1 and V-2 weapons, before the possible Soviet seizure of the former Nazi fabrication plants and potential hostage taking of scientists, engineers, and their families. In 1945, Maj. Gen. H.N. Toftoy requested the immediate transport of 300 Germans to the United States, initiating Operation Paperclip.

By May 18, 1948, Operation Paperclip had brought a total of 1,136 German scientists and engineers to the United States, along with 100 nearly complete V-2 rockets, 300 railroad boxcars of V-2 components, plans, manuals, and other documents. A select group of scientists and engineers were sent to Fort Bliss, Texas, and White Sands Proving Ground, the centers for rocket development until 1950.

Following World War II, the Army Ordnance Department guided missile program was transferred to the reactivated Redstone Arsenal on June 1, 1949, as the new center for research and development of rockets along with chemical and conventional ammunition storage.

In November 1950, some 130 German scientists, engineers, and supporting personnel moved to Redstone Arsenal. Between 1950 and 1952, the Redstone rocket was developed, the first large American ballistic missile, a direct descendant of the V-2 rocket. The Army Ballistic Missile Agency was established February 1, 1956. Rocket propulsion research with spaceflight applications gained momentum, particularly after the success of Sputnik on October 4, 1957. Four months later, the ABMA responded with Explorer 1 aboard a Jupiter-C (Juno) rocket in January 1958—America's first satellite. Afterwards, the Juno program, a derivative of the Redstone rocket, was renamed Saturn.

NASA was created on July 29, 1958, to manage the civilian space program. Its first center for propulsion, sited in the heart of Redstone Arsenal, was dedicated on September 8, 1950, and named in honor of Gen. George C. Marshall.

An A-3 rocket is in Test Stand 3 at Kummersdorf, Germany, south of Berlin. Test Stand 3 was not only mobile but also the largest stand. The Aggregat (A) series of rockets were the forerunners of the V-2 (Vergeltungswaffe 2 or "vengeance weapon") developed at Peenemunde Army Research Center. By 1943, Adolf Hitler deployed the V-2 as a "wonder weapon" to maintain morale in Germany. The Redstone rocket was the scion of the A series. (Courtesy of Library of Congress.)

On October 28, 1949, the secretary of the Army approved the transfer of the Ordnance Research and Development Division Sub-Office (Rocket) at Fort Bliss, Texas, to Redstone Arsenal as the Ordnance Guided Missile Center, later named the Army Ordnance Missile Command (AOMC) Headquarters. By 1958, it included the ABMA, Redstone Arsenal, Jet Propulsion Laboratory (JPL), White Sands Proving Grounds, and the Army Rocket and Guided Missile Agency. (Author's collection.)

Two technicians for Thiokol Corporation attend to the solid rocket propellant motor of an RV-A-10 missile, built by General Electric. Both companies collaborated to produce an efficient, long-duration rocket for the Hermes program that was successfully flight-tested in 1953. At 14 feet, 4 inches long and 31 inches in diameter, this was the technological forebear of the space shuttle solid rocket boosters. (Courtesy of the Huntsville–Madison County Public Library.)

Following a group photograph at Fort Bliss, Texas, in the late 1940s, where they worked on guided missile development, the German rocket team was transferred to the ABMA Redstone Arsenal in Huntsville in 1950. The team continued to work on guided missile development until transferring to the new George C. Marshall Space Flight Center. (Courtesy of NASA.)

A Nike Hercules surface-to-air missile stands guard outside the Army Ordnance Guided Missile School. Established in 1953 during the Cold War missile race, by 1960, it had trained 20,000 technicians from all branches of the US armed forces and 12 allied nations. It was renamed Army Ordnance Munitions and Electronic Maintenance School and moved to Fort Lee, Virginia, in 2011. (Courtesy of The Huntsville–Madison County Public Library.)

By 1954, the unique military-civilian team was rapidly pushing forward the research, design, development, production, maintenance, and supply of all US Army missiles and rockets. The command proclaimed people to be its primary resource and their brains and ability its most precious asset, requiring both military police and civilian guards to protect. (Author's collection.)

Cold War security concerns and low-profile, high-stakes research and development in the field of ballistic and guided missiles required a combination of military and civilian police to patrol the arsenal's southern border by boat along the Tennessee River, as seen here on March 3, 1954. (Author's collection.)

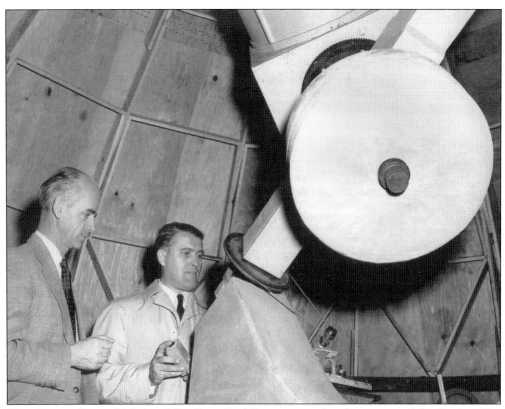

Wernher von Braun (right) and Dr. Ernst Stuhlinger are at the observatory of the Rocket City Astronomical Association in 1956 in Monte Sano State Park. The association was founded in 1954 by 16-year-old Sammy Pruitt and his classmates who requested the help of the missile and rocket scientists at the nearby arsenal. Over the years, it has upgraded its telescope and built a planetarium. In 1972, it was renamed the Von Braun Astronomical Society. (Author's collection.)

On November 11, 1954, eight scientists and two of their wives, brought to the United States from Germany via Operation Paperclip, became American citizens in Huntsville. The first groups of scientists arrived at Fort Bliss and White Sands Proving Grounds in late 1945 before being transferred to Redstone Arsenal on April 1, 1950. (Courtesy of The Huntsville–Madison County Public Library.)

Charles Lundquist stands at the blackboard explaining orbit theory and tracking for Project Orbiter to Hermann Oberth (left) and Wernher von Braun at the ABMA Redstone Arsenal on June 28, 1958. Project Orbiter (Army) was a competitor to Project Vanguard (Navy). It was rejected, but its basic design was used for the Juno I launch vehicle for Explorer 1, the United States' first satellite. (Courtesy of NASA.)

The arrival of ballistic and guided missiles, courtesy of the US Army, was widely regarded as a symbol of Huntsville's great progress and investment in the future. Holding a model Redstone rocket are members of the Huntsville Chamber of Commerce (from left to right) Walter Fleming, director; unidentified; M.H. Lanier Jr., president; Carl Woodall, outgoing president; and Jack Langhorne, vice president and director. (Author's collection.)

On February 22, 1956, a Redstone ballistic missile, with an unofficial range of 200-300 miles, is being loaded onto a launching platform by crane. A direct descendant of Germany's World War II–era V-2 rocket, the Redstone was one of the US Army's first steps toward its goal to develop a guided missile that could accurately carry an atomic warhead to targets 1,500 miles away. (Author's collection.)

German engineers met with Maj. Gen. Holger N. Toftoy (back left), commander of the US Army Ballistic Missile Agency and director of Operation Paperclip, at Redstone Arsenal in 1956. Gathered around the table are, from left to right, Ernst Stuhlinger, Hermann Oberth, Wernher von Braun, and Dr. Robert Lusser, who returned to Germany in 1959. (Courtesy of NASA.)

Huntsville's prolific rocket scientists and members of the military made good use of the abundant offerings found in the Huntsville Carnegie Library. Magazine selections ranged from the *New Yorker* and *Collier's* to *Scientific Monthly* and *Army Life*. Seated are William Fanning (left) and Alfred Smith accompanied by an unidentified dog and two patrons. (Courtesy of the Huntsville–Madison County Public Library.)

This photograph shows the Redstone Interim Test Stand (left) and instrumentation and control tanks (right). To the left of the rocket is the Cold Calibration Tower and an access platform to reach the top of the rocket. Developed by Wernher von Braun and associates with a budget of $25,000 in 1953, it was constructed out of materials salvaged from around the arsenal. The blockhouses to the right are railroad tank cars. (Courtesy of Library of Congress.)

Maj. Gen. John B. Medaris was greeted by Redstone Arsenal commander Brig. Gen. Holger N. Toftoy and a 13-gun salute when he arrived to become commanding general of the new Army Ballistic Missile Agency in late January 1956. During World War II, Medaris served as captain of the Army Ordnance Department, supporting the 1st Army, and identified the first German V-2 rocket fired against it in Europe. (Author's collection.)

With a Redstone missile locked in place, the 145-foot Static Test Tower awaits its next assignment. A static test consists of firing a missile held in place by the test stand, allowing its engine performance to be studied as if in flight. Completed in 1957 at a cost of $12 million, it was the largest static firing test stand for rocket motors in the free world, surpassing the Redstone Interim Test Stand. (Author's collection.)

With the growing success of missile research and development, culminating in the feasible development of the Redstone rocket and Explorer 1, this group of city boosters was promoting Huntsville as "Rocket City U.S.A." on the streets surrounding Courthouse Square in 1957. (Courtesy of Huntsville–Madison County Public Library.)

Maj. Gen. John B. Medaris (left), commander of Redstone Arsenal, and Maj. Gen. Holger N. Toftoy (right), commander of the US Army Ballistic Missile Agency, are accompanied by two children and a Hermes guided missile at the corner of Memorial Parkway and Airport Road in 1956. The event was the unveiling of a historical marker commemorating the Hermes as the first American-made guided missile put on public display, in 1953. (Courtesy of The Huntsville–Madison County Public Library.)

Less than two months before the birth of NASA, the members of the National Advisory Committee for Aeronautics Special Committee on Space Technology, called the Stever Committee, met at Lewis Research Center on May 26, 1958. Counterclockwise from right are Wernher von Braun, Abe Silverstein, Dale Corson, Hugh Dryden, H. Guilford Stever, Carl Palmer, J.R. Dempsey, Robert Gilruth, H. Julian Allen, Milton Clauser, Samuel Hoffman, W. Randolph Lovelace, Hendrik Bode, Abraham Hyatt, Col. Norman Appold, and Edward Sharp. (Author's collection.)

A 1958 news feature was allowed this photograph inside the Jupiter ballistic missile fabrication shop at Redstone Arsenal, which was usually veiled in secrecy. The two rockets at left have exposed liquid oxygen tanks while awaiting their engine components to be welded onto the flange. The Jupiter evolved into the Juno I launch vehicle that placed the first American satellite, Explorer 1, into orbit. (Author's collection.)

This graphic shows the components and science instruments aboard Explorer 1, the first US satellite that launched during the International Geophysical Year on January 31, 1958. After the Soviet Union's Sputnik on October 4, 1957, the ABMA was directed to launch a satellite using its Jupiter-C rocket as the launch vehicle. The JPL designed, built, and operated the payload in less than three months. (Author's collection.)

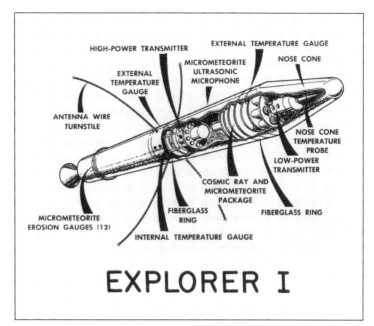

Shown with a model of the Jupiter-C rocket are, from left to right, (seated) Dr. Eberhard Rees, Gen. John B. Medaris, and Wernher von Braun; (standing) Ernst Stuhlinger and two unidentified. The rocket consists of a modified Redstone ballistic missile topped by three solid-propellant upper stages that used clusters of scaled-down rocket engines. When used as a launch vehicle, it is sometimes referred to as the Juno-I. (Author's collection.)

Residents of Huntsville celebrate in Courthouse Square on January 31, 1958, as news arrived that Juno I had successfully launched Explorer 1, the free world's first artificial satellite, both built at Redstone Arsenal. From left to right, Jimmy Walker, Mayor R.B. "Speck" Searcy, Stuart Jones, and Dorsey Uptain improvise a Molotov cocktail four weeks after the launch of Sputnik. (Courtesy of Huntsville–Madison County Public Library.)

Wernher von Braun was presented with a framed page of the *Huntsville Times* commemorating the success of America's first artificial satellite carrying a scientific research instrument. Designed by Dr. James van Allen of the University of Iowa, it was a cosmic ray detector to measure the radiation environment in Earth orbit. Subsequent satellites confirmed that a belt of charged particles was trapped by Earth's magnetic field, known as the Van Allen Belt. (Author's collection.)

Production specialist Val Stapler gives a lively guided tour of the inner workings of a Jupiter rocket on November 10, 1958. Manufactured by prime contractor Chrysler Corporation, the triangular brace is removed to allow the engine to be slipped into one of three rails or slots (seen just above Stapler's outstretched hand) inside the aft section of the launch vehicle. (Author's collection.)

Members of the Huntsville Chamber of Commerce and their guests tour Redstone Arsenal and the fabrication and assembly facility for the Jupiter rockets on May 27, 1959. Originally designed as an intermediate range ballistic missile, the Jupiter missile evolved into the Juno family of space launch vehicles. On January 31, 1958, it launched the free world's first satellite, Explorer 1. (Courtesy of Huntsville–Madison County Public Library.)

A lone soldier stands guard with the watchdog of the skies, a Nike missile, in firing position at Redstone Arsenal, on June 1, 1958. First built in 1945 by the Douglas Aircraft Company with a radar system designed by Bell Laboratories, the Nike series was designed for defense against attack by high-flying bombers or ballistic reentry vehicles. (Author's collection.)

On August 20, 1958, Pres. Dwight D. Eisenhower (center) swears in Dr. T. Keith Glennan (right), as NASA's first administrator, and Dr. Hugh L. Dryden as its deputy administrator. The National Aeronautics and Space Act established the civilian space program on July 29, 1958. NASA replaced the National Advisory Committee for Aeronautics, established March 3, 1915, less than 12 years after the Wright brothers' first flight. (Author's collection.)

President Eisenhower and Katherine Marshall, wife of George C. Marshall, unveil a red granite bust of Gen. George C. Marshall sculpted by Finnish sculptor Kalervo Kallio during a brief, quiet ceremony on September 8, 1960. From left to right are T. Keith Glennan, NASA administrator; Wernher von Braun, director of MSFC; Eisenhower; Marshall; and Maj. Gen. August Schomburg, commanding general of AOMC. (Courtesy of the Huntsville–Madison County Public Library.)

Following the dedication on September 8, 1960, von Braun (center, right) escorted President Eisenhower (center, left) and other dignitaries on a tour of the site, including a 1959 Saturn I scale mock-up, composed of Redstone and Jupiter tanks. The space complex within the boundaries of Redstone Arsenal became MSFC by executive order on March 15, 1960. (Courtesy of the Dwight D. Eisenhower Presidential Library and Museum.)

From left to right, von Braun, Eisenhower, and Karl L. Heimburg tour the test facilities at the newly named MSFC on September 8, 1960. As test laboratory director, Heimburg told his engineers that they were not only responsible and accountable for test operations, but also safety, quality, and reliability of the test stand, test vehicle, and test operations. (Courtesy of the Dwight D. Eisenhower Presidential Library and Museum.)

At the dedication of the George C. Marshall Space Flight Center on September 8, 1960, President Eisenhower (front row, center, holding hard hat) and Wernher von Braun (next to Eisenhower) are surrounded by civilian and military dignitaries. It was deemed that cooperation between NASA and the Department of Defense should continue, but civilian space exploration was now NASA's responsibility. (Courtesy of the Dwight D. Eisenhower Presidential Library and Museum.)

By the end of 1960, Huntsville had become the home of Redstone Arsenal, the development center for rockets and guided missiles, and the George C. Marshall Space Flight Center. It had seen the establishment of the University of Alabama at Huntsville, the Rocket City Astronomical Association, a symphony orchestra, a population increase of 23 percent from the previous decade, and widespread recognition as a major center for high technology in a span of 10 short years. (Author's collection.)

Five

THE COLD WAR SPACE RACE AND APOLLO
1960–1973

In 1960, NASA described its new field installation as "the only self-contained organization in the nation which was capable of conducting the development of a space vehicle from the conception of the idea, through production of hardware, testing, and launching operations." Even the initial design for the launch complex in Florida was performed at MSFC; engineers traveled there to conduct launch activities and then back to Huntsville to analyze the data.

Its goals were to reach space, know and understand the space environment, and inhabit and utilize space for the benefit of mankind.

The federal government wanted to ensure that the United States did not fall behind its Communist rival, the Soviet Union. But on April 12, 1961, Soviet cosmonaut Yuri Gagarin became the first man to orbit the Earth.

On April 25, 1961, Pres. John F. Kennedy began a dramatic expansion of the US space program, placing the vice president at the head of NASA. He then addressed Congress on May 25 on the subject of urgent national needs, requesting additional funding and proclaiming that "this nation should commit itself to achieving the goal, before the decade is out, of landing a man on the Moon, and returning him safely to Earth." On July 20, 1969, Apollo 11 astronauts Neil Armstrong, Michael Collins, and Edwin "Buzz" Aldrin Jr. became the first of several Americans to do so.

The major endeavors of MSFC were space vehicles, space science, and manned systems. Its notable achievements, while numerous, included S-IB, the free world's first manned spacecraft; Saturn V, the world's largest launch vehicle; LM, the world's only manned lunar landing vehicle; Skylab, the free world's first space station with the nation's largest orbital observatories; and Spacelab, the first materials processing experiments in space, site of the first commercial product made in space and the development of the propulsion systems for the world's first space shuttle.

Shortly after the unprecedented success of Apollo 11, Pres. Richard M. Nixon announced deep budget cuts for NASA, making its next program strictly Earth orbital in scope.

Following the success of Sputnik on October 4, 1957, President Eisenhower called for a long-term, reasoned, and integrated space program, as well as the production of a "Super Jupiter" in the short run. In 1959, a version of the Saturn I was fabricated from Redstone and Jupiter tanks. "The need made us more inventive," said ABMA planner Willi Mrazek. (Courtesy of NASA.)

Developed under the direction of MSFC by North American Aviation's Rocketdyne division, a cluster of H-1 engines for the Saturn I launch vehicle is in the alignment fixture. Use of a cluster of propellant tanks from Redstone and Jupiter missiles sitting on top of a single thrust plate and then attaching the engines to the bottom of the plate greatly reduced development time of the S-I. (Courtesy of NASA.)

Military and civilian personnel watch as a Saturn I first-stage booster rocket endures a 122-second test in the Static Test Tower (S-I/I-B), Building 4572, East Test Area, with a Juno missile in the tower's west position. The S-I's eight H-1 engines generated 1.3 million pounds of thrust on June 17, 1960. To the left is Building 4514, Liquid Hydrogen Facility (S-IV-B). (Author's collection.)

The urgency and importance of MSFC's mission in the 1960s was apparent from the beginning. On April 12, 1961, the *Huntsville Times* reported what Wernher von Braun called "the shot heard around the world"—the Cold War adversary of the United States had sent the first human into space. (Courtesy of NASA.)

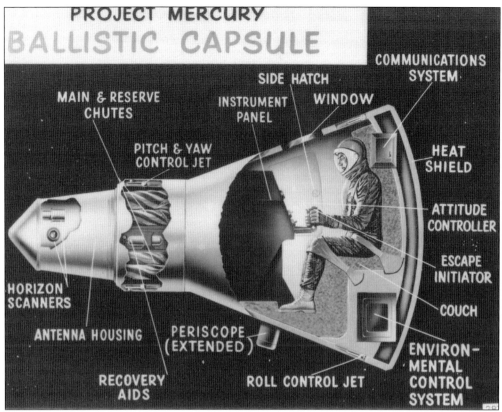

PROJECT MERCURY
BALLISTIC CAPSULE

COMMUNICATIONS SYSTEM

SIDE HATCH

WINDOW

MAIN & RESERVE CHUTES

INSTRUMENT PANEL

HEAT SHIELD

PITCH & YAW CONTROL JET

ATTITUDE CONTROLLER

ESCAPE INITIATOR

HORIZON SCANNERS

COUCH

ANTENNA HOUSING

PERISCOPE (EXTENDED)

RECOVERY AIDS

ROLL CONTROL JET

ENVIRON- MENTAL CONTROL SYSTEM

This cutaway drawing of a Mercury capsule was used by the Space Task Group at the first NASA inspection on October 24, 1959. The blunt-shape body was the creation of Dr. Maxime A. Faget, the designer of the Mercury spacecraft. In 1962, he became director of engineering and development at the Manned Spacecraft Center. Project Mercury proved that humans could live and work in space, paving the way for all future human exploration. (Author's collection.)

On January 1, 1960, technicians install a Mercury capsule and escape system atop a Redstone launch vehicle prior to a static test firing at MSFC's Redstone Test Stand. The Mercury-Redstone launch vehicle was designed to place a manned capsule in suborbital flight and allow for subsequent safe recovery. (Courtesy of NASA.)

On October 27, 1961, NASA marked a high point in the three-year-old Saturn development program when the first Saturn vehicle flew a flawless 215 mile ballistic trajectory from Cape Canaveral, Florida. The SA-1 is pictured in the Static Test Tower (S-I/S-IB), Building 4572, five months before launch on May 16, 1961, following a proof test of internal pressure only or internal pressure plus extreme loads on the propellant tanks. (Courtesy of NASA.)

The Saturn I launch vehicle (S-I) is being placed into Building 4557, Dynamic Test Stand (Saturn I/IB), in the East Test Area on June 1, 1961. The booster was vertically mated for the first time with dummy S-IV and S-V stages. The assembled vehicle was then readied for dynamic testing to investigate the integrity of the support structure. (Courtesy of NASA.)

Pres. John F. Kennedy (right) was accompanied by, from left to right, Wernher von Braun, NASA associate administrator Dr. Robert C. Seamans Jr., Vice Pres. Lyndon B. Johnson, and NASA administrator James E. Webb, as they walk past a dome structure with one gore segment on display for inspection on September 11, 1962. (Courtesy of Robert Knudsen White House Photographs, John F. Kennedy Presidential Library and Museum, Boston, Massachusetts.)

Surrounded by onlookers, President Kennedy greets an MSFC technician at the historic Redstone Test Site. Turned away from the camera are Wernher von Braun (left) and NASA administrator James E. Webb. (Courtesy of Robert Knudsen White House Photographs, John F. Kennedy Presidential Library and Museum, Boston, Massachusetts.)

A static test fire of a Saturn rocket booster at MSFC was also a demonstration of American technological might. President Kennedy's visit was part of a two-day inspection tour of NASA field installations that also included Cape Canaveral, the Manned Spacecraft Center in Houston, and McDonnell Aircraft Corporation in St. Louis. (Courtesy of Cecil Stoughton White House Photographs, John F. Kennedy Presidential Library and Museum, Boston, Massachusetts.)

On the last day of his two-day tour of NASA field installations, Kennedy persuasively set forth the parameters and goals of America's space program at the stadium of Rice University near Houston on September 22, 1962. The goal of landing a human on the Moon and returning safely to Earth by the end of the decade was accomplished on July 20, 1969. (Author's collection.)

Visitors pass through the Rocket Garden before entering the Space Orientation Center. The rockets at left are, from front to back, V-2, Hermes, Jupiter, Redstone, Juno II, Jupiter-C, and Mercury-Redstone. At far right is the Saturn I S-I (first stage), which flew on October 27, 1961. (Courtesy of NASA.)

The Space Orientation Center displayed the many benefits of America's space program. On December 28, 1964, a 1961 Ford Country Squire station wagon is parked outside the main entrance. In the center's brochure, von Braun explained that "a continuing mandate from the people will allow America to pursue a sound space program worthy of our imagination and ideals, permitting us to use this knowledge for the benefit of all mankind." (Courtesy of NASA.)

On October 5, 1964, Evelyn Falkowski, curator of the Space Orientation Center (SOC), presents a mock-up of the Saturn I Instrument Unit (IU) to Israeli colonel Amos Horev. After moving to Huntsville in 1958 with her husband, an electrical engineer at Redstone Arsenal, Falkowski, a lawyer, became active in civil rights after accepting a position as historian at the SOC. (Courtesy of NASA.)

Three Saturn I S-Is are in varying states of assembly inside the Fabrication and Assembly Engineering Laboratory, Building 4705, on January 13, 1963. Destined for use in the SA-4, SA-6, and SA-7 missions, the S-I consisted of clusters of 1.78-meter tanks from the Redstone rocket and 2.67-meter tanks from the Jupiter missiles. Since the engines would be clustered, so would the tanks. (Courtesy of NASA.)

A technician sets up radiometer equipment for an RL-10 engine for MSFC at Lewis Research Center on December 12, 1963. The Saturn I S-IV (second stage) was powered by six RL-10 engines, built by North American Aviation Rocketdyne Division. It was the first large liquid hydrogen and oxygen rocket engine to be built in the United States. (Courtesy of NASA.)

Technicians position a spider beam that forms the forward structure of the stage and serves to anchor the forward end of the propellant containers. Seal plates then cover the forward side of the spider beam. A liquid oxygen/solid oxygen (LOX/SOX) disposal system is installed above the seal plates to purge the S-IV interstage area during S-IV engine chill down. (Courtesy of NASA.)

A Saturn I S-I (first stage) is in transit to the Static Test Tower (S-I/S-IB), Building 4572, in the East Test Area. Eight Saturn Is were designed and built at MSFC; two were built by Chrysler Corporation at MSFC. Chrysler was a "level-of-effort" contractor that provided manufacturing know-how and labor to build a government design. (Courtesy of NASA.)

On October 1, 1967, the flight version of the Saturn IB launch vehicle's first stage arrives at MSFC's Static Test Tower (S-I/S-IB). Between December 1967 and April 1968, the stage would undergo seven static test firings. The S-IB, developed by MSFC and built by Chrysler Corporation at Michoud Assembly Facility in New Orleans, utilized eight H-1 engines, each producing 200,000 pounds of thrust. (Courtesy of NASA.)

From left to right, Dr. Karl L. Heimberg, Wernher von Braun, and Maj. Gen. John B. Medaris examine a model of the Propulsion and Structural Test Facility. Built in 1957 by the Army Ballistic Missile Agency, it was transferred to NASA MSFC as the primary center for development of large launch vehicles and rocket propulsion systems in 1958. A framed photograph of a Nike-Hercules ballistic missile is on the wall. (Author's collection.)

Project Gemini principals tour the NASA Manned Spacecraft Center in Houston on June 22, 1964. From left to right are Dr. George Mueller, NASA associate administrator; Maxime A. Faget, director of engineering and development; and Charles Matthews, manager of Project Gemini. Gemini had four goals: test an astronaut's ability to fly long-duration missions, rendezvous and dock spacecraft in orbit, perfect reentry and landing methods, and understand the effects of longer spaceflights. (Author's collection.)

Fritz Pauli, known as a "one-man rocket company," examines his scale model of the Saturn SA-5. He specialized in small-scale versions of the H-1 and F-1 engines to better examine potential problems and testing. With degrees in electrical and mechanical engineering, he was an authority on rocket propulsion and was granted nine patents during his career. Pauli rejoined von Braun's team in Huntsville in 1952. (Author's collection.)

The Saturn V Dynamic Test Stand, also known as the Dynamic Structural Test Facility, was built in 1964 to conduct mechanical and vibrational tests on the fully assembled Saturn V Space Vehicle. Standing 360 feet high and 122 feet by 98 feet at the base, it could test each rocket component separately as well as in partial and full assembly. It was modified for the space shuttle and remains on standby today. (Author's collection.)

An S-IB-200D, a dynamic test version of the Saturn IB launch vehicle's first stage, is hoisted into Building 4557, Dynamic Test Stand, on January 11, 1965, to test its structural soundness. Employing a "building block" approach to Saturn rocket development, the uprated Saturn IB utilized Saturn I technology to further develop and refine booster size and capabilities for eventual manned lunar missions. (Courtesy of NASA.)

In the Static Test Tower (S-1/S-1B), Building 4572, is a Saturn S-IB first-stage booster undergoing a test fire with a Juno II in the west position of the stand. The uprated S-IB was built by Chrysler Corporation at the Michoud Assembly Facility in New Orleans and tested at MSFC. Five S-IBs supported Project Apollo, three orbited Skylab crews, and one boosted the Apollo/Soyuz Test Project mission. (Courtesy of NASA.)

Larger facilities were required for static testing of the colossal Saturn V S-IC booster rocket and its five F-1 engines. Construction of the Saturn Static Test Stand (S-IC) also included related facilities, such as the blockhouse that served as the control center for the test stand. The two were connected by a narrow access tunnel that also housed the control cables, shown under construction on June 13, 1962. (Courtesy of NASA.)

The S-IC Static Test Stand, Building 4670, utilized hundreds of tons of steel and 12 million pounds of concrete, planted down to bedrock 40 feet below ground level. The foundation walls are 4 feet thick, while the base structure has four towers with 40-foot-thick walls extending upward 144 feet above the ground, topped by a 135-foot boom, making it among the tallest structures in Alabama in 1962. (Courtesy of NASA.)

On October 1, 1963, technicians are nearing completion of a full-scale mock-up of a Saturn V S-IC first-stage thrust structure together with F-1 engines in the Manufacturing/Engineering Laboratory. The use of mock-up prototypes enabled design and size evaluation, in addition to overall fit. (Courtesy of NASA.)

On December 1, 1964, Jeanette Scissum-Mickens, a master of science in mathematics, works at her desk. Note the scale model of a Jupiter-C and the US Air Force lunar wall mosaic on the wall—the target of the Apollo program. She worked in the Space Sciences Laboratory on the Atmospheric, Magnetospheric, and Plasmas in Space Project. Her published papers covered everything from solar prediction models to equal employment opportunities. (Courtesy of NASA.)

Karl L. Heimburg, director of the Astronautics Laboratory (far left), talks with Wernher von Braun and Walt Disney as they tour the West Test Area, especially the new S-IC Static Test Stand, Building 4670, with a Saturn V S-IC in place. In 1965, von Braun hoped to reignite Disney's enthusiasm for space exploration in films similar to the live action and animated features they had collaborated on a decade earlier. (Author's collection.)

From left to right, George C. Wallace, governor of Alabama; James Webb, NASA administrator; and Wernher von Braun, director of MSFC, are pictured during a tour of MSFC on June 8, 1965. Governor Wallace and Dr. Webb, along with members of the Alabama state legislature and press reporters, were there to witness the first test firing of a Saturn V booster rocket. (Author's collection.)

A technician is overseeing the welding of a Y-ring to the S-IC bulkhead and fuel tank on March 1, 1965. The Y-ring was designed to eliminate lap joints where the tank dome wall and adjoining structure, such as the intertank segment, come together. This procedure was for the Saturn VSA-502 launch vehicle, an uncrewed Saturn V test, and final uncrewed Apollo test mission. (Courtesy of NASA.)

In late 1964, the fuel tank assembly for kerosene (RP-1) and thrust structure of the Saturn V S-IC (first stage) are readied to be mated to the LOX tank. MSFC engineers produced the general design, while Boeing manufactured it at the Michoud Assembly Facility. (Courtesy of NASA.)

On December 1, 1964, the fuel tank assembly for propellants RP-1 (left) and LOX are mated in Building 4705, Assembly Shop and Hangar. Combined, these propellants fed the five F-1 engines attached to the S-IC first stage that lifted the entire Saturn V vehicle from the launch pad. (Courtesy of NASA.)

An array of F-1 engines are being prepared for shipment on March 1, 1965. Developed by North American Aviation's Rocketdyne Division under the direction of MSFC, the F-1 was the most powerful single-nozzle liquid-fuel rocket engine ever flown. A cluster of five F-1 engines mounted on the Saturn V S-IC burned 15 tons of LOX and RP-1 per second to produce 7.5 million pounds of thrust. (Courtesy of NASA.)

Leaving the Manufacturing and Engineering Laboratory in 1965, the Saturn V S-IC-T was en route to the newly built S-IC Static Test Stand, Building 4670, in the West Test Area. Known as the S-IC-T, the booster was a static test vehicle not intended for flight. It was ground tested repeatedly over a period of many months, proving the vehicle's propulsion system. (Courtesy of NASA.)

Engineers and technicians are installing one of five F-1 engines into the thrust structure of an S-IC first-stage booster after the booster itself was installed into the Saturn Static Test Stand in 1965. The F-1 engine, measuring 19 feet tall by 12.5 feet in diameter at the nozzle exit, was installed on the booster after it had been placed in the test stand. (Courtesy of NASA.)

The Saturn V S1-C-T, a propulsion test article not intended for flight, is being hoisted into the S-IC Static Test Stand, Building 4670, in 1965. By April 1966, all S-IC static test operations were performed at the newly completed, dual-position B-1/B-2 Test Stand at the Mississippi Test Facility (MTF). (Courtesy of NASA.)

John C. Houbolt explains his concept for lunar landings that would insert a spacecraft into lunar orbit while a smaller lander would descend to the Moon and then rendezvous with the orbiting craft to return to Earth. The competitor concept, known as "direct ascent," required a mega-size launch vehicle called Nova. Lunar orbital rendezvous was deemed to be the most efficient method of landing Apollo spacecraft on the Moon. (Author's collection.)

Pictured at center, a Lunar Roving Vehicle (LRV), developed and tested at MSFC, wanders the simulator area in 1965. Also known as the "Moon Buggy," it provided astronauts greater mobility on the lunar surface. In the upper right is the 18-acre Random Motion/Liftoff Simulator, or "Arm Farm," developed to test the Saturn swinger mechanisms used to hold the rocket in position until liftoff. (Courtesy of NASA.)

The "Arm Farm" was a unique 18-acre facility capable of testing the detachment and reconnection of at least six arm test positions under virtually realistic conditions. Each position had two elements that duplicated motions during countdown and launch using umbilical connections and personnel access hatches driven by a hydraulic servo system to replicate motion between the vehicle and the tower. (Courtesy of NASA.)

Wernher von Braun, accompanied by Mayor Joe Davis of Huntsville and Madison County Commission chairman James Record (both to his right, from left to right), addresses the large crowd gathered on the steps of the county courthouse and Courthouse Square to celebrate the tremendous success of the Apollo 11 mission that placed the first human on the Moon and returned him safely to Earth in July 1969. (Courtesy of the Huntsville–Madison County Public Library.)

Sitting at the head of the table is Wernher von Braun; his son Peter and wife, Maria, are sitting on the near side of the table opposite him. They are relaxing with friends and well-wishers at the MSFC picnic ground on July 26, 1969. The brochure "Lunar Landing Celebration, Picnic-Open House-Dance" described the occasion as an event "to commemorate the successful participation by the Marshall Space Flight Center in man's first expedition to the surface of the Moon." (Author's collection.)

Our deep appreciation for your outstanding contribution to the success of Apollo 11

George E. Mueller Sam C. Phillips Kurt H. Debus Robert R. Gilruth Wernher von Braun

This commemorative photograph of the principals of the Apollo program, originally signed and inscribed by Gen. Samuel C. Phillips, Apollo program director, was subsequently signed by, from left to right, George Mueller, NASA associate administrator; Gen. Sam Phillips; Kurt H. Debus, director of Kennedy Space Center (KSC); Robert R. Gilruth, director of the Manned Space Center; and Wernher von Braun, director of MSFC. (Courtesy of NASA.)

In early 1970, the thermal unit that controlled the temperature stability of the Apollo Telescope Mount (ATM) was installed into a vacuum chamber. Designed and developed by MSFC, it was a solar observatory aboard Skylab that was manually operated by the astronauts from 1973 to 1974. During space walks, or extravehicular activities (EVAs), astronauts climbed to the circular disk on top to replace camera film magazines, which they took back to Earth. (Courtesy of NASA.)

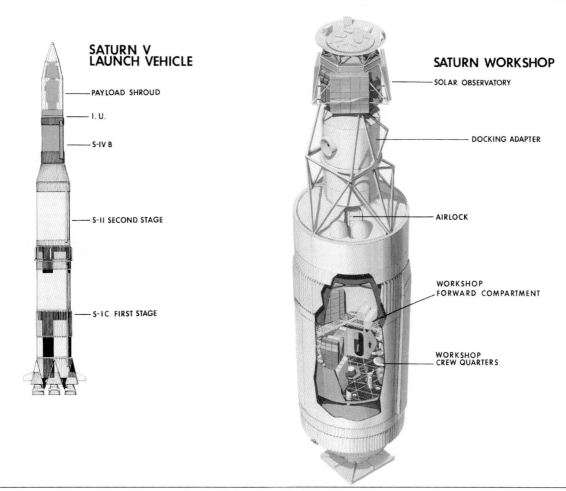

SATURN V
LAUNCH VEHICLE

- PAYLOAD SHROUD
- I. U.
- S-IV B
- S-II SECOND STAGE
- S-IC FIRST STAGE

SATURN WORKSHOP

- SOLAR OBSERVATORY
- DOCKING ADAPTER
- AIRLOCK
- WORKSHOP FORWARD COMPARTMENT
- WORKSHOP CREW QUARTERS

MSFC developed and integrated most of the major Skylab components utilizing a payload shroud to protect the Saturn workshop, which used a Saturn V launch vehicle for its last flight. The three successive crewed launches used Saturn IB launch vehicles. McDonnell Douglas Corporation was contracted to convert an S-IV (third stage), previously used for the Saturn V, into the orbital workshop with more comfortable crew quarters. (Courtesy of NASA.)

Amid dwindling budgets following the successful Moon landing of Apollo 11 on July 20, 1969, NASA transferred Wernher von Braun to its headquarters in Washington, DC, to oversee strategic planning. In February 1970, crowds gathered around the Madison County Courthouse to bid "Auf Wiedersehen" to him and his family, seated to the right of the podium on the steps of the courthouse. The banner in the background reads, "Huntsville's first citizen. . . . on loan to Washington, D.C." (Courtesy of Huntsville–Madison County Public Library.)

Six

STS AND THE
SPACE SHUTTLE
1973–2011

On January 5, 1972, Pres. Richard M. Nixon and NASA administrator James C. Fletcher announced a new approach to space exploration that would build on the successes of Apollo, including Skylab and the Apollo-Soyuz Test Project. The space shuttle program would take "the astronomical costs out of astronautics" and would provide routine access to space and cost savings in the form of a reusable vehicle.

With the Apollo lunar landing program ending on December 7, 1972, NASA launched four Skylab missions beginning on May 14, 1973, through February 1974. MSFC supplied the Skylab workshop itself, plus the four Saturn launch vehicles, the solar observatory, and many scientific experiments for each mission. It began an era of comprehensive scientific research in space. An MSFC Saturn rocket was used the last time as part of the joint US-Soviet Apollo-Soyuz Test Project mission in 1975.

With the final shuttle configuration selected, MSFC was responsible for the development of its advanced propulsion systems. The principal shuttle elements—orbiter, space shuttle main engines (SSME), external tank (ET), and solid rocket boosters (SRB)—were all developed under Marshall Center management except the orbiter itself.

The orbiter served as both passenger and cargo vehicle and required highly efficient propulsion systems. The SSMEs are the most advanced cryogenic liquid-propellant rocket engines ever built and will be used for the future Artemis project. The ET was the propellant tank and the structural backbone of the entire shuttle assembly; it had to be strong and lightweight, its design simple, and cost minimal since it was not reusable. The SRBs were reusable and were the first solid-propellant rockets built for a manned space vehicle and the largest ever flown.

The world's first reusable space vehicle, Space Transportation System-1 (STS-1), including the orbiter *Columbia*, was successfully launched on April 12, 1981, and returned to Earth on April 14, 1981.

During the shuttle period, changes in NASA's philosophy and resources, principally the issue of reuse and increased awareness of limited resources, challenged MSFC to become a leaner, stronger institution as it adapted to these changes.

SKYLAB
PRIMARY MODE
PARASOL THERMAL SHIELD
DEPLOYED THRU SAL

Skylab 1 launched on May 14, 1973. During ascent, it lost its thermal protection shield and debris held a solar array wing slightly open, making the workshop uninhabitable. A crewed Skylab 2 repair mission, launched May 25, attached a parasol to the Scientific Airlock (SAL) and extended the folded shield through the opening and into space. An EVA freed the jammed solar array, and the Skylab program was saved. (Courtesy of NASA.)

Two seamstresses work on the parasol sunshade that "fix-it" Skylab II astronauts Charles Conrad Jr., Paul Weitz, and Joseph Kerwin brought to the Skylab Orbital Workshop (OWS). The sunshade was composed of three layers: aluminized mylar, laminated ripstop nylon, and thin nylon. NASA and contractor personnel at Johnson Space Center, KSC, and MSFC collaborated on the emergency repair procedure. (Courtesy of NASA.)

On May 20, 1973, a technician checks procedures to be used to raise the sunshade that will be used to cool the crippled orbiting Skylab. In the Neutral Buoyancy Simulator (NBS) that holds a submerged mock-up of Skylab, the technician is at the central work position of the ATM, where he controls the lines that will raise and spread the shield from the bag in the foreground. (Author's collection.)

Surrounded by trash bags, two Skylab 4 astronauts, science pilot Edward G. Gibson (left) and Comdr. Gerald P. Carr, peer into the airlock module looking the length of the Skylab OWS. Not shown is William Pogue, pilot. They were the third and final crew aboard Skylab for a record-breaking 24 days in space, parlaying a day off from communicating with Mission Control for a more self-directed, reduced workload. (Courtesy of NASA.)

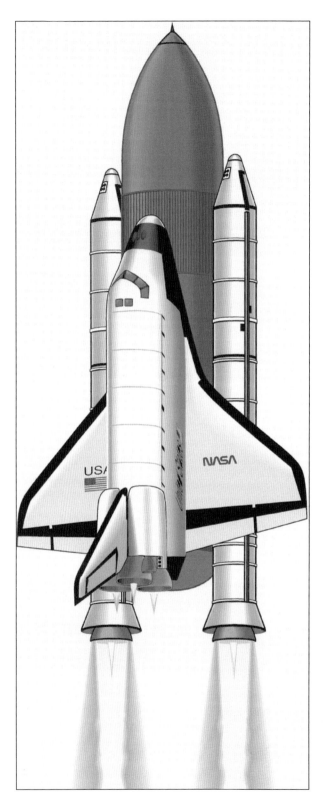

The space shuttle was the world's first reusable spacecraft, and MSFC played a leading role in the design, development, testing, and fabrication of many major shuttle propulsion components. Of the three major components, MSFC developed the orbiter's high-performance SSMEs, the massive ET, and the SRBs. The space shuttle program began an era of sustained science in low-Earth orbit. (Courtesy of NASA.)

On August 1, 1977, a technician reams holes to the proper size and alignment in the main injector body of the RS-25 SSME, through which propellants will pass on their way into the engine's combustion chamber. The SSME is a cryogenic liquid-propellant rocket engine that produces 418,000 pounds of thrust at liftoff. It is manufactured by Rocketdyne, a division of Rockwell International, under contract to MSFC. (Courtesy of NASA.)

An SRB structural test article is in place in Building 4572 at MSFC for structural and load verification tests on November 11, 1978. At nearly 150 feet high, the twin boosters provide 5.8 million pounds of thrust at liftoff. The SRBs were the largest solid-fuel rockets ever built and the first designed for recovery, refurbishment, and reuse. (Courtesy of NASA.)

An early ET is installed into Building 4670 (Static Test Stand) on March 1, 1978. The largest component of the STS, the ET is also the structural backbone of the system, providing support for attachment with the SRBs and orbiter. At liftoff, the ET absorbs the total thrust loads of 7.8 million pounds of the three SSMEs and the two SRBs. (Courtesy of NASA.)

A LOX tank, part of the shuttle's ET, undergoes a hydroelastic modal survey test at MSFC on March 1, 1978. Along with the liquid hydrogen tank and the intertank, the ET provides propellants to the shuttle's three main engines during the first eight and a half minutes of flight. It was manufactured by the Martin Marietta Corporation at Michoud Assembly Facility. (Courtesy of NASA.)

Technicians lower the nose cone into place to complete stacking of the left SRB inside Building 4557 (Dynamic Test Stand) in the East Test Area on September 1, 1978. The platforms had been modified to accommodate the two SRBs to which the ET would be attached. Later, the orbiter would be attached to the ET for the Mated Vertical Ground Vibration Test (MVGVT) in October 1978. (Courtesy of NASA.)

This was the view from the second-floor platform, looking up at subsequent platforms, of the Saturn V Dynamic Structural Test Facility, Building 4550, in the East Test Area. The platform edges outline the modifications made to the test stand to accommodate the space shuttle components. (Courtesy of the Library of Congress.)

The space shuttle orbiter *Enterprise* is removed from the Dynamic Test Stand, Building 4550, following its first MVGVT on March 1978. It was the first time the shuttle components were mated vertically. The MVGVT provided an experimental base in the form of structural dynamic characteristics for the shuttle vehicle that were then compared to the predicted analytical results. (Courtesy of NASA.)

An engineer at MSFC observes a model of the space shuttle components being tested in the 14-by-14-inch Trisonic Wind Tunnel, which is capable of subsonic, transonic, and supersonic air speeds. The tunnel operates by high-pressure air flowing from storage to either vacuum or atmospheric conditions. It has been part of the space program since 1958. (Courtesy of NASA.)

On June 7, 1981, astronaut Bruce McCandless (STS-31, STS-41), pictured at center, works with a full-scale mock-up of a portion of the space telescope that was designed to be serviced on-orbit in MSFC's NBS. A water environment that allows practice in simulated weightlessness, the NBS allows repair and component change procedures in zero gravity to be documented ahead of a shuttle launch. (Courtesy of NASA.)

Two astronauts accompanied by divers practice construction techniques to build the International Space Station (ISS) in the NBS, Building 4705, at MSFC on December 1, 1985. The NBS is 75 feet in diameter, 40 feet deep, and contains 1.3 million gallons of water. It provides a weightless environment for testing hardware for use in space. (Courtesy of NASA.)

Designed by MSFC, the Long Duration Exposure Facility (LDEF) launched aboard the orbiter *Challenger* STS-41-C on April 6, 1984, carried 57 science and technology experiments designed by more than 200 investigators. Some of MSFC's experiments included trapped proton energy determination, linear energy transfer spectrum measurement, and solar array materials. The LDEF was retrieved in January 1990 by *Columbia* STS-32 and brought back to Earth for analysis. (Courtesy of NASA.)

Spacelab was a reusable laboratory developed by the European Space Agency for certain flights flown by the space shuttle allowing scientists to perform a variety of experiments in microgravity. There were 22 major Spacelab missions between 1983 and 1998 alone. MSFC managed the Spacelab missions aboard the space shuttle at the Huntsville Operations Support Center and the Spacelab Payload Operations Control Center. (Courtesy of NASA.)

Pictured in 1991, Dr. Gerald Fishman, an astrophysicist at MSFC and principal investigator of the Compton Gamma-Ray Observatory (GRO) instrument, works on the the Burst and Transient Source Experiment (BATSE) detector module, which would alert scientists to gamma-ray bursts. Designed and built by MSFC, it launched aboard *Atlantis* STS-35 in April 1991 and ended a successful nine-year mission in June 2000. (Courtesy of NASA.)

The Compton GRO is being deployed by the Remote Manipulator System arm aboard *Atlantis* STS-37 in April 1991. It served as a gamma-ray alert for the Hubble Space Telescope, the Chandra X-Ray Observatory, and ground-based observatories around the world. A total of 37 universities, observatories, and NASA centers in 19 states, as well as 11 institutions in Europe and Russia, participated in the BATSE science program. (Courtesy of NASA.)

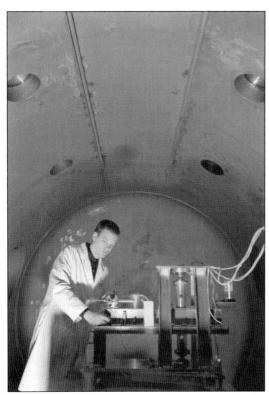

On June 20, 1996, engineers at one of MSFC's vacuum chambers began testing a microthruster model that would allow NASA to develop microthrusters to move the space shuttle, a future space station, or any other space-related vehicle with the least amount of expended energy. In space, gravity is so minuscule that a very small force can move very large objects. Microthrusters are used to produce these small forces. (Courtesy of NASA.)

The Chandra X-ray Observatory High Resolution Mirror Assembly is being removed from the test structure in the X-Ray Calibration Facility (XRCF) at MSFC. The most sophisticated and powerful x-ray telescope ever built is managed by MSFC, whose XRCF is the largest, most advanced laboratory for simulating x-ray emissions from distant celestial objects. It was launched July 22, 1999, aboard *Columbia* (STS-93). (Courtesy of NASA.)

The Huntsville Simulation Laboratory (HSL) is a facility that tests and verifies the SSME avionics and software system using a maximum complement of flight-type hardware. The HSL permits evaluations and analyses of the SSME avionics hardware, software, control system, and mathematical models. It also performs a wide variety of tests and verified operational procedures to ensure component compatibility under diverse operating conditions. (Courtesy of NASA.)

The joint airlock module of the ISS is undergoing exhaustive structural and systems testing in the space station manufacturing facility at MSFC prior to shipment to KSC. The airlock, 18 feet long and with a mass of approximately 13,500 pounds, was launched to the ISS aboard *Atlantis* STS-104 on July 12, 2001. MSFC played a primary role in the development, manufacturing, and operations of the ISS. (Courtesy of NASA.)

A 24-inch version of the space shuttle's reusable solid rocket motor was successfully fired for 21 seconds at a MSFC test stand; this test-fire was done to ensure that a replacement insulating material met MSFC's ongoing verification of components, materials, and manufacturing processes. Directed by MSFC and manufactured by ATK Thiokol Propulsion Division, each motor generates an average thrust of 2.6 million pounds during its two-minute burn. (Courtesy of NASA.)

Standing before the orbiter *Atlantis* outside the Orbiter Processing Facility at KSC are astronauts (from left to right) Robert Crippen, STS-1 pilot; Charles Bolden, NASA administrator, STS-31, -45, -60, and -61-C; Janet Kavandi, STS-91, -99, and -104; Robert Cabana, KSC director, STS-41, -53, -65, and -88; and Mike Parrish, *Endeavour* vehicle manager for United Space Alliance. Major General Bolden announced the permanent display locations for the four orbiters at the end of the space shuttle program. (Courtesy of NASA.)

Seven

DEEP SPACE EXPLORATION AND ARTEMIS

2012–PRESENT

Previously identified by its individual components, NASA's current lunar program was designated project Artemis in May 2019 and will serve as a collective umbrella for the development of the Space Launch System (SLS), a new heavy-lift rocket, and the Orion Multi-Purpose Crew Vehicle (MPCV), which includes the European Service Module (ESM) and the Lunar Orbital Platform-Gateway (LOP-G). In 2017, Space Policy Directive 1 steered NASA toward a 2024 crewed mission to the Moon's south pole. Hardware and services will be developed by NASA, along with domestic and international private aerospace companies, under its Commercial Lunar Payload Services.

The ambitious plan to carry the next man and the first woman to the Moon will use the SLS heavy-lift rocket to send the four-person Orion MPCV and ESM to the Gateway. Astronauts will then continue to the Moon's surface via lunar landers developed by commercial aerospace companies such as Astrobotic, Intuitive Machines, and OrbitBeyond.

While the SLS, Orion MPCV, and ESM are all in varying stages of development and testing, the priority is to develop and launch the Gateway's Power and Propulsion Element and a utilization module, a small habitat module. Due to budget constraints, a continuously crewed multi-module Gateway orbiting the Moon will be delayed until 2028. Each element of the Gateway will be launched commercially as it is developed.

A series of increasingly complex missions will begin with Artemis 1, formerly known as Exploration Mission-1 (EM-1), an uncrewed launch around the Moon in 2020–2021. Artemis 2 will carry a crewed launch around the Moon in 2023, and Artemis 3 will conclude by landing humans on the Moon via the Gateway and lunar lander in 2024.

According to Greek mythology, Apollo and Artemis were the twin offspring of Zeus (an Olympian) and Leto (a Titan, descendant of the deities who preceded the Olympians). Zeus ruled as king of the gods on Mount Olympus. Artemis became the goddess of the Moon and the hunt; her companion and fellow hunter was Orion.

Home to the SLS program, MSFC's Building 4220 is a blend of aesthetics, practicality, and high efficiency. It features state-of-the-art green technologies and energy conservation systems that include low-emissive glass, rooftop solar-power units, a 10,000-gallon cistern to collect stormwater, and a bio-swale that captures and filters stormwater runoff and directs it to a collecting pond. It is the seventh LEED-certified structure at MSFC. (Courtesy of NASA.)

An Orion MPCV test article, attached to the Launch Abort System (LAS), is hoisted up the Vertical Integration Facility at Space Launch Complex 46 at Cape Canaveral, where it will be stacked atop a booster for an Ascent Abort-2 (AA-2) flight test on July 2, 2019. The LAS will pull the Orion crew capsule away from the SLS rocket in the event of an emergency during ascent. (Courtesy of NASA.)

The Orion MPCV, slated for Artemis 1, is pictured after it underwent a Direct Field Acoustics Test (DFAT), which exposed it to maximum acoustic levels of 141 decibels (dB). The DFAT was performed at NASA's Plum Brook Station testing facility. A remote, 6,400-acre site that is part of the Glenn Research Center in Sandusky, Ohio, it hosts four world-class test facilities for complex and innovative ground testing. (Courtesy of NASA.)

The Orion stage adapter structural test article for NASA's Artemis SLS makes its way past Building 4200 at MSFC to the Redstone Army Airfield for delivery to Lockheed Martin in Denver on July 10, 2017. There, it will be loaded into NASA's Super Guppy, the current iteration of a stretched cargo plane built by Aero Spacelines/Airbus and capable of accommodating wide, heavy loads. (Courtesy of NASA.)

Members of Leadership Alabama Inc., Anthony Davis (left), Alabama house minority leader, and Dr. Stephen Leahy, Auburn University president, pause before Test Stand 4693 on March 7, 2019, with an Artemis SLS core stage in place. Built on the foundation of the F-1 Engine Test Stand 4696, at 215 feet tall, the test stand contains hydraulic cylinders that duplicate the same loads and stresses produced during launch. (Courtesy of NASA.)

New Orleans mayor Mitchell J. Landrieu (left, foreground), Louisiana governor John Bel Edwards (third from left), and two onlookers listen intently to MSFC director Todd May discuss the deep space launch vehicle being built by Boeing at Michoud Assembly Facility. (Courtesy of NASA.)

Boeing's SLS liquid hydrogen tank test article is moved onto the barge *Pegasus* at the MAF dock on December 14, 2018. Originally designed and built to transport the space shuttle ET, *Pegasus* was enlarged from 260 feet to 310 feet and qualified to carry more than 600,000 pounds to transport the SLS from MAF to Stennis Space Center and MSFC. (Courtesy of NASA.)

The LOX tank structural test article of the SLS core stage, manufactured by Boeing at MAF, is moved into the Vertical Assembly Building, where simulators that mimic the intertank and the forward skirt are connected to the tank. It will then be carried by the *Pegasus* barge to MSFC for structural testing. (Courtesy of NASA.)

The LOX tank structural test article is lifted into Test Stand 4697 at MSFC on July 10, 2019. The LOX tank and liquid hydrogen (LH2) are the two propellant tanks inside the core stage that will produce more than two million pounds of thrust that will launch Artemis 1 to the Moon. In the test stand, dozens of hydraulic cylinders will simulate liftoff and flight stresses. (Courtesy of NASA.)

Steve Doering (left), manager of the stages element of the SLS program, explains the dimensions and capabilities of the full-stack scale model of the Artemis SLS. In the lobby of Building 4200, Central Laboratory and Office Building at MSFC, are Sen. G. Douglas Jones and his wife, Louise New Jones. They were touring the entire facility and test stands. (Courtesy of NASA.)

Pictured in the Systems Integration Laboratory (SIL), developer George Plattsmier, who helps support hardware and software systems, integration, and testing for the Artemis Space Launch System, is surrounded by the tools of his trade in Lab 116. SIL is part of Building 4205, Propulsion Research and Development Laboratory, which is currently designing, building, and testing the avionics for Artemis. (Courtesy of NASA.)

A prototype lunar lander, Peregrine, by Astrobotic was on display at Goddard Space Flight Center on May 31, 2019. Under the Commercial Lunar Payload Services program, it is one of nine proposals from commercial aerospace companies to provide transportation between the Artemis Gateway to the lunar surface and back. Nova-C by Intuitive Machines, Z-01 by OrbitBeyond, and Astrobotic are the first three commercial landers selected for the program. (Courtesy of NASA.)

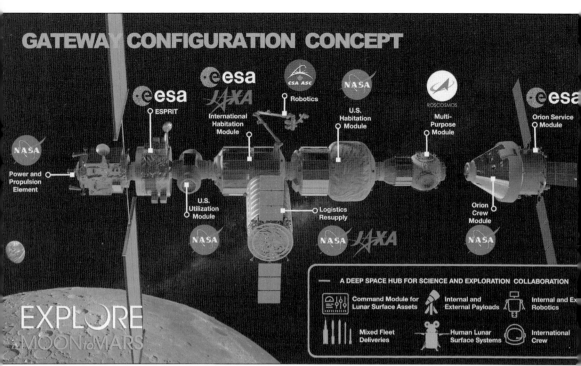

The Artemis LOP-G is a modular space station that will support crewed missions, science requirements, and technology demonstrations. Governed by the Power and Propulsion Element, the solar-powered hub will provide a secure area for lunar landers that will ferry astronauts to the Moon and back. It will also serve as the command and communications center of the Gateway. (Courtesy of NASA.)

Eight

GEN. GEORGE C. MARSHALL
1880–1959

The youngest son of a coal merchant, George Catlett Marshall was born in Uniontown, Pennsylvania, 50 miles southeast of Pittsburgh. Graduating from Virginia Military Institute in 1901, he was known for his quiet self-confidence and self-discipline, as well as his talent for communicating among diverse groups.

Commissioned as a second lieutenant in 1902, he was an aide-de-camp to General Pershing by 1918, becoming US Army chief of staff in 1939. During World War II, he organized the largest ground and air force in the history of the United States. British prime minister Winston Churchill called him "the organizer of victory." Pres. Franklin D. Roosevelt considered General Marshall so valuable he remarked that "he could not sleep at ease if he were out of Washington." In 1944, the president's chief of staff became a five-star general of the Army.

Appointed secretary of state by Pres. Harry S. Truman, General Marshall proposed and organized the strategies that became the European Recovery Program. Known as the Marshall Plan, it rebuilt Europe in the aftermath of World War II, providing economic aid to those countries that could become vulnerable to Communist overtures. In recognition, Marshall became the first general of the Army to receive the Nobel Peace Prize, in 1953.

Following Marshall's death, Pres. Dwight D. Eisenhower announced that NASA's propulsion development and research center would be named in his honor. Commenting on the president's selection in 1959, Dr. Hugh L. Dryden, NASA deputy administrator, wrote, "The new center comes to NASA from the Army and its personnel have long been associated with the Army. It seems fitting therefore, to honor a great military leader whose life was dedicated to the cause of peace, who initiated the Marshall Plan, who won the Nobel Peace Prize, and who served our country as Secretary of State. The bestowal of his name on the new center would not only honor his character, his leadership, and his dedication to public service, but also serve as a reminder to those associated with the laboratory to follow his example."

General Marshall is pictured in 1940, one year after being appointed Army chief of staff (1939–1945) by President Roosevelt. After World War II, he was named special ambassador to China, served as secretary of state and president of the Red Cross, and was awarded the Nobel Peace Prize for the European Recovery Program, or Marshall Plan. (Courtesy of the Library of Congress.)

The Marshall-Cole families gathered for a photograph before a wedding ceremony in Lexington, Virginia. From left to right are Marie Marshall Singer; Elizabeth "Lily" C. Coles, the bride, George C. Marshall Jr., the groom; Stuart Marshall; Laura and George Marshall Sr.; and Lizzy P. Coles. On February 11, 1902, after a brief reception in the parlor of the bride's home, she declared, "Come on, George, let's get married." (Courtesy of the George C. Marshall Foundation, Lexington, Virginia.)

Sitting in the fourth row, second from left, 1st Lt. George C. Marshall was a member of the US Army School of the Line, class of 1907, at Fort Leavenworth, Kansas. Established in 1881 by William Tecumseh Sherman as a college and training school for infantry and cavalry officers, it was renamed School of the Line in 1907. (Courtesy of the George C. Marshall Foundation, Lexington Virginia.)

Aide-de-camp George C. Marshall (right) rides in parade before the Arc de Triomphe in Paris with General of the Armies John J. Pershing (center) in 1919, shortly after the end of World War I on November 11, 1918. Following graduation from Virginia Military Institute in 1901, Marshall was stationed in the Philippines during the insurrection and became a five-star general by 1944. (Courtesy of the Library of Congress.)

From left to right, Brig. Gen. Frank Parker, commander of the 1st Division; Col. James A. Drain, former member of the 1st Division staff; and Lt. Col. George C. Marshall, 1st Division operations officer, are on the steps of the White House after meeting with the president. The group had requested the honor of Pres. Calvin Coolidge speaking at the dedication of the 1st Division Monument on October 4, 1924, commemorating its members who fought in World War I. (Courtesy of the Library of Congress.)

On November 27, 1939, General Marshall (left), the newly appointed US Army chief of staff, and Rep. Edward T. Taylor (right) appeared before the House Appropriations Subcommittee in Congress in support of President Roosevelt's request for $271,999,523 to finance national defense deficiencies. A large number of Americans favored isolationist policies to stay out of the war in Europe that had begun on September 1, 1939. (Courtesy of the Library of Congress.)

At the National Aviation Forum on May 27, 1940, General Marshall (center), the US Army chief of staff, confers with Thomas Morgan (left), president of Sperry Gyroscope Corporation, and Thomas Beck, president of Crowell Publishing Company. Marshall explained that more than emotion was needed to produce 50,000 warplanes a year; world war aviation experience indicated "the need for careful, calm, and coordinated planning." (Courtesy of the Library of Congress.)

Maj. Gen. George C. Marshall (left, standing) was a close aide to President Roosevelt (seated) during World War II. Here, Brig. Gen. William H. Wilbur (kneeling) is receiving the Congressional Medal of Honor in the field at Casablanca, Morocco, from Maj. Gen. George S. Patton, with the president and chief of staff in attendance. (Courtesy of the Franklin D. Roosevelt Presidential Library and Museum.)

In December 1943, the leaders of the Allied powers met in Tehran, Iran, during World War II. From left to right are an unidentified British officer; US Army chief of staff General Marshall; Sir Archibald Clark Keer, British ambassador to the Soviet Union; Harry Hopkins, secretary of commerce and President Roosevelt's closest advisor on foreign policy; Marshal Stalin's interpreter; Marshal Joseph Stalin; foreign minister Vyacheslav Molotov; and Gen. Kliment Voroshilov. (Courtesy of the Library of Congress.)

Chief of staff Marshall (left), with Gen. Henry H. Arnold, is in attendance at the Second Quebec Conference (code-named "Octagon") on September 12–16, 1944. It was the second high-level military conference held between British and American governments during World War II; the first was "Quadrant" in 1943. (Courtesy of the Franklin D. Roosevelt Presidential Library and Museum.)

Gen. Dwight D. Eisenhower (left) greets well-wishers in the back of a jeep accompanied by the chief of staff Gen. George C. Marshall, at the airport in Washington, DC, on June 18, 1945. Victory in Europe had been declared on May 8, with the formal acceptance by the Allied powers of Nazi Germany's unconditional surrender. (Courtesy of the Dwight D. Eisenhower Presidential Library and Museum.)

Marshall (right) passes on parade during the Potsdam Conference from July through August 2, 1945, in occupied Germany. The Allied leaders, Prime Minister Winston Churchill, Pres. Harry S. Truman, and Marshal Joseph Stalin, met to establish postwar order, address peace treaty issues, and methods to counter the effects of World War II. (Courtesy of the Harry S. Truman Presidential Library and Museum.)

Marshall (left) is sworn in by Chief Justice Fred M. Vinson in the Oval Office as President Truman watches on January 8, 1947. A few of the distinguished guests are, from left to right, Sen. Arthur Vandenberg, Adm. William Leahy, Gen. Harry Vaughn, Secretary of the Treasury John Snyder, and Attorney General Tom Clark. (Courtesy of the Harry S. Truman Presidential Library and Museum.

A worker in Germany helps lay the foundation for one of several low-cost housing projects financed by the Marshall Plan. The banner on the building proclaims, "Emergency program for Berlin with the Marshall-Plan-Aid." (Courtesy of the George C. Marshall Foundation, Lexington, Virginia.)

From left to right, President Truman; Secretary of State Marshall; Paul G. Hoffman, economic cooperation administrator; and Ambassador W. Averell Harriman discuss the Marshall Plan in the Oval Office on November 29, 1948. Officially named the European Recovery Program, it became effective April 3, 1948, to provide economic assistance to Europe after World War II. (Courtesy of the Harry S. Truman Presidential Library and Museum.)

Former Secretary of State George C. Marshall (right) is presented a commemorative album titled "The Marshall Plan at Mid Mark" in April 1950. The presenters are Secretary of State Dean Acheson (left), who helped design the Marshall Plan, and Paul G. Hoffman, who led the implementation of the Marshall Plan between 1948 and 1950. (Courtesy of the Harry S. Truman Presidential Library and Museum.)

Discover Thousands of Local History Books
Featuring Millions of Vintage Images

Arcadia Publishing, the leading local history publisher in the United States, is committed to making history accessible and meaningful through publishing books that celebrate and preserve the heritage of America's people and places.

Find more books like this at
www.arcadiapublishing.com

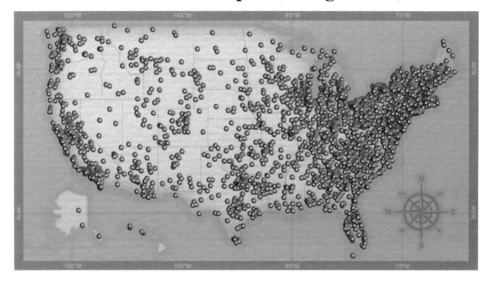

Search for your hometown history, your old stomping grounds, and even your favorite sports team.

Consistent with our mission to preserve history on a local level, this book was printed in South Carolina on American-made paper and manufactured entirely in the United States. Products carrying the accredited Forest Stewardship Council (FSC) label are printed on 100 percent FSC-certified paper.

MADE IN THE